AF361686

Energy-Efficiency of Conveyor Belts in Raw Materials Industry

Energy-Efficiency of Conveyor Belts in Raw Materials Industry

Editors

Daniela Marasová
Monika Hardygora
Mirosław Bajda

MDPI • Basel • Beijing • Wuhan • Barcelona • Belgrade • Manchester • Tokyo • Cluj • Tianjin

Editors

Daniela Marasová
Technical University of Košice
Slovakia

Monika Hardygora
Wroclaw University of
Science and Technology
Poland

Mirosław Bajda
Wroclaw University of
Science and Technology
Poland

Editorial Office
MDPI
St. Alban-Anlage 66
4052 Basel, Switzerland

This is a reprint of articles from the Special Issue published online in the open access journal *Energies* (ISSN 1996-1073) (available at: https://www.mdpi.com/journal/energies/special_issues/Energy_ Efficiency_of_Conveyor_Belts).

For citation purposes, cite each article independently as indicated on the article page online and as indicated below:

LastName, A.A.; LastName, B.B.; LastName, C.C. Article Title. *Journal Name* **Year**, *Volume Number*, Page Range.

ISBN 978-3-0365-4385-7 (Hbk)
ISBN 978-3-0365-4386-4 (PDF)

Contents

Editorial

Energy Efficiency of Conveyor Belts in Raw Materials Industry

Mirosław Bajda [1,*,†], Monika Hardygóra [1,†] and Daniela Marasová [2,†]

[1]   Faculty of Geoengineering, Mining and Geology, Wroclaw University of Science and Technology, 50-421 Wrocław, Poland; monika.hardygora@pwr.edu.pl

[2]   Faculty of Mining, Ecology, Process Control and Geotechnology, Technical University of Kosice, 042 00 Kosice, Slovakia; daniela.marasova@tuke.sk

*   Correspondence: miroslaw.bajda@pwr.edu.pl; Tel.: +48-71-320-68-92

†   These authors contributed equally to this work.

Citation: Bajda, M.; Hardygóra, M.; Marasová, D. Energy Efficiency of Conveyor Belts in Raw Materials Industry. *Energies* **2022**, *15*, 3080. https://doi.org/10.3390/en15093080

Received: 27 March 2022
Accepted: 20 April 2022
Published: 22 April 2022

Publisher's Note: MDPI stays neutral with regard to jurisdictional claims in published maps and institutional affiliations.

1. Introduction—Special Issue Information

Belt conveyors are presently the most common transportation machines used in surface and underground mining. Apart from the mining industry, belt conveyors are also used in cement and chemical industries, as well as in power plants and ports. Conveyor belt transportation technology used in global mining industry has undergone constant development. The advances are most clearly visible in long-distance belt conveyor designs. The two most important areas for belt conveyor development include first and foremost improving belt conveyor efficiency and extending the length of a single conveyor, which entails increasing the power of the drive mechanisms. High-power conveyors need considerable amounts of electric energy, and consequently, due to global increase in energy prices, they generate increased transportation costs. In recent years, very intensive research has been performed on lowering energy consumption of belt conveyor drive mechanisms. Extensive theoretical and experimental research demonstrated the potential for energy savings in individual components of belt conveyors, such as belts, idlers, gearboxes, couplings, drive systems, belt tensioning systems, etc. Identification of main resistance in belt conveyors is an example of one such research approach. The research proved that the feasibility of limiting electric energy consumption is linked to belt properties. Estimations suggest that implementing improved energy-saving belt with adequate parameters of rubber cover will allow for a significant decrease in conveyor primary resistance, which will result in decreased electric energy consumption by conveyor drive mechanisms.

Analysis of belt transportation systems only in Polish brown coal mines shows the scale of the problem. The "Belchatow" lignite mine, which extracts above 4×10^7 Mg of coal and more than 1×10^8 m^3 of overlay per year, may be a good example here. The transportation of materials in the "Belchatow" mine is performed with the use of belt conveyors having a total length of more than 160 km and accounts for approximately 50% of its electric energy consumption. This fact demonstrates both the economic and ecological importance of technologically optimizing belt conveyors in order to lower the energy consumption of drive mechanisms.

The proposal of topics for the Special Issue of the journal *Energies* includes the following research areas:

1. Energy-saving solutions in belt conveyor transportation—modeling of the operating conditions and dimensioning of conveyors, monitoring the condition of conveyor components, and predictive diagnostics.
2. New energy-saving solutions with respect to conveyor components, especially their drive systems, gearboxes, couplings, idlers, as well as devices for controlling and monitoring their operation.
3. Calculations of belt conveyor parameters and of energy savings due to conveyor speed adjustments, more efficient use of the transportation capacity, and optimal selection of conveyors for particular tasks.

4. Applications of environmentally friendly and economically justified design solutions aimed at improving the energy efficiency of belt conveyors and limiting their noise impact.
5. Optimizing the operation-related processes in belt conveyors (evaluation of the energy consumption, the durability, and the reliability of conveyor transportation systems).
6. Methods for the evaluation and measuring of both the quality of conveyors and their behavior during the starting process.
7. Experiments regarding the laboratory and in-service tests of belt conveyors and their components, such as belts, idlers, drives and gearboxes—new measurement and result-processing technologies.
8. The use of belt conveyors related to the monitoring of their condition, damage analysis, computer-aided management, as well as the identification of the properties of both, belts and their splices.

In the next section, we provide a brief overview of the papers published in the Special Issue according to the previously outlined thematic areas.

2. A Short Review of the Contributions in This Issue

Results of tests into the energy efficiency of belt conveyor transportation systems indicate that the energy consumption of their drive mechanisms can be limited by lowering the main resistances in the conveyor. The main component of these resistances is represented by belt indentation rolling resistance. Limiting its value will allow for a reduction in the amount of energy consumed by the drive mechanisms. Bajda and Hardygóra [1] present a test rig which enables uncomplicated evaluations of such rolling resistances. The rolling resistance tests on the test stand, which the authors called the "inclined plane", can be used to search for the optimal parameters of rubber compounds and optimal design solution for belt bottom covers. The article also presents the results of comparative tests performed for five steel-cord conveyor belts. The tests involved a standard belt, a refurbished belt, and three energy-saving belts. As temperature significantly influences the values of belt indentation rolling resistance, the tests were performed in both positive and negative temperatures. The results indicate that when compared with the standard belt, the refurbished and the energy-efficient belts generate higher and lower indentation rolling resistances, respectively. In order to demonstrate practical advantages resulting from the use of energy-saving belts, this article also includes calculations of the power demand of a conveyor drive mechanism during one calendar year, as measured on a belt conveyor operated in a mine. The replacement of a standard belt with a refurbished belt generates a power demand that is higher, and with an energy-efficient belt it is lower. For this reason, the authors suggest that refurbished conveyor belts should be withdrawn from service.

Olchówka et al. [2] compares two methods of identification and classification of steel cord failures in the conveyor belt core based on an analysis of a two-dimensional image of magnetic field changes recorded using the DiagBelt system around scanned failures in the test belt. One of the aspects undertaken in this study is the statistical analysis of the obtained data, which allows determining the correlation between the input data and the class to which the damage belongs. These values were compared with the values determined for the test set. It is, however, worth noting that with such a choice of analysis for automatic recognition of damage, it is necessary to execute many damage measurements for multiple sets of parameters to obtain a sample that comes from the same distribution as the training data. This solution can be cumbersome and, as the study shows, verification by the mean of the sample data is not always reliable. In the second stage of the research, artificial intelligence methods were applied to construct a multilayer neural network (MLP) and to teach it appropriate identification of damage. The failure classification using a neural network showed high efficiency. In both methods, the same data sets were used, which made it possible to compare methods. It is worth noting that while analysis of statistical methods has already been used in the classification of belts damage, cluster analysis and analysis using the neural networks have so far been rarely discussed and their results rarely presented.

Doroszuk et al. [3] addresses the conveyor transfer station design problem in harsh operating conditions. Conditions include very slopped receiving conveyor belts, acute angle between conveyors, and demanding atmospheric conditions. The work aims to identify and eliminate a failure phenomenon that interrupts aggregate supply. The failure—blockage of the transfer station—is observed during the transfer of the high volume of the material. The analyzed transfer station is located in a Polish granite quarry "Graniczna" in Dolnośląskie Voivodship. The first stage of the study employs laser scanning of the transfer station and the feeding and receiving conveyor belts. The reverse engineering methods used the point cloud aggregated from 3D scans to map the existing transfer station and its geometry. CAD model created from the point cloud was prepared so that the geometry of the transfer station and conveyors could have been represented by flat surfaces that had contact with bulk material. Next, a discrete element method (DEM) model of granite aggregate has been prepared. The preparation of the DEM model involved parametrization of the granite from the quarry, including Young modulus, Poisson ratio, density, coefficients of restitution, and rolling and sliding friction. The tested parameters were calibrated. Using calibration box in which bulk material was flowing from the upper part to the lower part, and the angles of repose were measured as the parameter for calibration. Parameters of the simulated material in the DEM simulation of the performed experiment were changed in small ranges to obtain the repose angle with the same accuracy as in an actual experiment. The final calibrated parameters were used together with the CAD model of the transfer station to simulate the current operating conditions. The arch formation has been identified as the main reason for breakdowns. The opening of the transfer station was too narrow, and when a large number of big rocks have been falling into the transfer station at the same time, there was a higher risk of forming the arch from rocks that were blocking the opening and causing the material to accumulate in the station and cause failure. Alternative design solutions for transfer stations were tested in DEM simulations. The most uncomplicated design for manufacturing incorporated an impact plate, and a straight chute has been selected as the best solution. The selected design eliminates geometry with the opening, thus is eliminating the risk of arch formation. The study also involved identifying areas of the new station most exposed to wear phenomena. A new transfer point was implemented in the quarry and resolved the problem of blockages, and the areas most exposed to wear were covered with special liners.

Bortnowski et al. [4] deals with the issues of transverse vibrations of the conveyor belt and the description of vibrations by models. The authors reviewed the literature on vibration models in static and dynamic conditions, attention was paid to the issue of tape movement. Models describing vibrations were analyzed, comparing them to results measured in laboratory conditions. A string model and a beam model were proposed, introducing a parameter describing the movement and speed of the belt called the equivalent force in the belt. The research was performed at a laboratory station, the following parameters were controlled: linear speed and belt tension of the spacing of idler supports. For frequency measurements, the SVAN979 noise and vibration meter by Svantek ltd and the NI USB-4432 measurement system with LabView software were used. The course of the research procedure, the conditions of experiments and the procedure of data from signal registration, through spectral analysis to obtaining vibration frequencies, were discussed. The results are presented in tabular form and on graphs, which allows the authors to assess the impact of individual parameters on the measured frequencies. In the next section, the calculation results were compared based on the proposed models with the measured frequencies. The results are compiled on a graph, where you can see which of the models better reflects reality, and that the spacing of the idlers can have a large impact on the frequencies obtained. The MAE was calculated for the string model and the beam model, the trend was set in the function of the spacing of the idler supports. The results show that for small spacing, smaller than 1.6 m, the measured frequency is better presented by the beam model. For spacing larger than 1.6 m, the string model becomes more optimal. The beam model is therefore more applicable in vibration analyses in the upper belt, while the string model is more applicable in vibration analyses in the return belt.

The aim of the experimental studies presented by Marasova et al. [5] was to investigate and compare the dynamic loading of a P2500 type textile-rubber conveyor belt. The dynamic loads occurred when the belt was struck by a specially designed punch, which simulated the impact of excavated material on the belt. A comparison was made between the relative energy absorption values obtained with and without the use of the belt support system. The experiments simulated various real-life conditions such as impact of materials with different masses and impact from different heights. On the basis of the conducted research, a model of the dependence of the relative amount of absorbed energy on selected parameters was created. Some parameters of the impact process were taken into account in the proposed dynamic model. The obtained regression model confirmed that the mass of the punch hitting the belt and the presence of the belt support system are the parameters having statistically significant effect on the amount of energy absorbed by the belt during the impact of the striker against it.

Dąbek et al. [6] presented automated methods of industrial inspection are becoming more popular each day, not only in the inspection of products, but also very demanding task of inspecting the workplace in search of any signs of damage to the equipment or safety related problems. Regarding this aspect, mines are no different from any other industrial facility, and new technologies emerge every day to keep people safe and to keep the production on track. Inspection tasks can be usually divided into several different, smaller problems that can be solved separately. One of them is the detection of faulty idlers—passive elements of a conveyor belt that allow for an extracted material to be automatically transported horizontally over long distances. Transportation process in mines can take place under unhealthy or even unsafe conditions for maintenance staff, therefore use of mobile robots, such as unmanned ground vehicles (UGV), is helpful in minimizing risks that have to be taken to keep the production going. Data collected from sensors mounted on proposed robots can have many forms, such as 3D scans, noise or image recording, just to name a few. In this case, infrared images were used in order to detect the faulty idlers, as they generate heat due to friction coming either from the damage of the internal bearing, or from the friction between the belt and the idler when the idler is completely stuck and does not rotate. The process is supported by pre-processing performed on the RGB images, that allows to focus on the belt conveyor by excluding the unnecessary parts of the video, that may contain noise affecting the results of implemented algorithm. The experiment has been performed on dataset collected in real industrial conditions by UGV, where the proposed algorithm has proven to be effective.

In mining industry, both surface and underground, belt conveyors are widely used for the horizontal transportation of materials. Automated inspection of such structure, especially from the point of view of the inspection of idlers, is a very complex problem, especially due to three aspects. The first problem is scale—it is difficult to measure every parameter at every point when the length of the conveyor can in some cases exceed 1 km. The second problem is accessibility—it is impossible to place sensors in every place that one would like to, especially for our case, it is impossible to put vibration sensors directly on the bearing of an idler. The third problem is environmental accessibility—especially in underground mining the environmental conditions (temperature, humidity, dustiness, etc.) do not allow for prolonged inspection activities performed by human, especially with a lot of measurement equipment that they would have to carry around. To address those issues, Shiri et al. [7] propose to use an inspection robot that (besides other functionality) is able to record acoustic signals. Using a mobile robot (either remotely driven or autonomous) can largely improve safety of the inspection tasks, as well as accuracy and repeatability of the data acquisition itself. Acoustic data analysis is already widely used in condition monitoring. The robot can record a short segment of acoustic signal next to each idler. Then, signal processing techniques are used for signal pre-processing and analysis to check the condition of the idler. The authors have determined that even if it is possible to identify the damage signature in such signal, there is a lot of disturbances, such as random noises or belt joint passing over the idlers. In such case, classical signal processing techniques cannot be used, and very often they indicate a fault even if a given idler is in a good condition. In this paper. the authors propose novel processing solutions that allow for proper analysis of the acquired signals.

The selection of a conveyor belt capable of performing a particular transportation task when installed on a belt conveyor is primarily informed by the tensile strength of this belt. It is intended to ensure that the forces in the operating belt will not lead to it breaking, i.e., to an event that is dangerous to both the personnel and to the conveyor. Typically, in the belt conveyor design process, the values of safety factors are empirically identified. Breaks in the so-called "continuous" (unspliced) belt sections, and not in the spliced areas, are infrequent but do happen in practice. Woźniak and Hardygóra [8] present some aspects that may account for such breaks in conveyor belts. It indicates the so-called "sensitive points" in design, especially in the transition section of the conveyor belt and in identifying the actual strength of the belt. The presented results include the influence of the width of a belt specimen on the identified belt tensile strength. An increase in the specimen width entails a decrease in the belt strength. The research involved develops in universal theoretical model of the belt on a transition section of a troughed conveyor in which, in the case of steel-cord belts, the belt is composed of cords and layers of rubber, and in the case of a textile belt, of narrow strips. The article also describes geometrical forces in the transition section of the belt and an illustrative analysis of loads acting on the belt. Attention was also devoted to the influence of the belt type on the non-uniform character of loads in the transition section of the conveyor. The transition section of the conveyor is a region in which the forces in the belt cross-section vary significantly. Such non-uniformity of belt loads was confirmed in the analysis performed with the use of an original theoretical model of the belt in the transition section of the troughed conveyor, which allows for the elastic properties of the belt and for the interaction with adjacent cables/strips. The lowest load non-uniformity was observed in the belt with multi-ply polyamide core, and higher values—in belts with polyester–polyamide cores, with solid woven cores, with aramid cores, and with steel-cord cores, respectively. When replacing a worn belt, each decision to change the belt type should be preceded by a new analysis of the geometrical parameters of the transition section in the belt conveyor In the case when the operator decides to install a belt having different elastic properties but without modifying the geometrical parameters of the transition section, the belt load nonuniformity may be increased by as much as several hundred percent. This fact is of special importance in the case of the transition section before the drive pulley, where typically, the highest force levels in the belt are observed.

3. Conclusions

The need for investigations into lower energy consumption of belt conveyor drive mechanisms is demonstrated most importantly, by the popularity of belt conveyors in the global mining industry. The articles presented in this Special Issue have created a set of solutions that may be of interest to researchers dealing with the energy efficiency of belt conveyors in mine transport. The material is also a source of information for those who deal with conveyor transport on a daily basis.

The solutions presented in the Special Issue have an impact on optimizing and thus reducing the costs of energy consumption by belt conveyors. This is due, inter alia, to the use of better materials for conveyor belts, which reduce its rolling resistance and noise, and also improve the ability to adsorb the impact energy coming from the material falling on the belt. The use of mobile robots designed to detect defects in the conveyor's components makes the conveyor operation safer and the conveyor works longer and there are no unplanned stops due to damage.

Further development of conveyor transport should aim at greater reduction of electric energy consumption by conveyor drives. Solutions presented in the discussed articles lead to this goal.

Funding: This research received no external funding.

Conflicts of Interest: The authors declare no conflict of interest.

References

1. Bajda, M.; Hardygóra, M. Analysis of the Influence of the Type of Belt on the Energy Consumption of Transport Processes in a Belt Conveyor. *Energies* **2021**, *14*, 6180. [CrossRef]
2. Olchówka, D.; Rzeszowska, A.; Jurdziak, L.; Błażej, R. Statistical Analysis and Neural Network in Detecting Steel Cord Failures in Conveyor Belts. *Energies* **2021**, *14*, 3081. [CrossRef]
3. Doroszuk, B.; Król, R.; Wajs, J. Simple Design Solution for Harsh Operating Conditions: Redesign of Conveyor Transfer Station with Reverse Engineering and DEM Simulations. *Energies* **2021**, *14*, 4008. [CrossRef]
4. Bortnowski, P.; Gładysiewicz, L.; Król, R.; Ozdoba, M. Models of Transverse Vibration in Conveyor Belt—Investigation and Analysis. *Energies* **2021**, *14*, 4153. [CrossRef]
5. Marasova, D.; Andrejiova, M.; Grincova, A. Dynamic Model of Impact Energy Absorption by a Conveyor Belt in Interaction with the Support System. *Energies* **2022**, *15*, 64. [CrossRef]
6. Dabek, P.; Szrek, J.; Zimroz, R.; Wodecki, J. An Automatic Procedure for Overheated Idler Detection in Belt Conveyors Using Fusion of Infrared and RGB Images Acquired during UGV Robot Inspection. *Energies* **2022**, *15*, 601. [CrossRef]
7. Shiri, H.; Wodecki, J.; Ziętek, B.; Zimroz, R. Inspection Robotic UGV Platform and the Procedure for an Acoustic Signal-Based Fault Detection in Belt Conveyor Idler. *Energies* **2021**, *14*, 7646. [CrossRef]
8. Woźniak, D.; Hardygóra, M. Aspects of Selecting Appropriate Conveyor Belt Strength. *Energies* **2021**, *14*, 6018. [CrossRef]

Article

An Automatic Procedure for Overheated Idler Detection in Belt Conveyors Using Fusion of Infrared and RGB Images Acquired during UGV Robot Inspection

Przemyslaw Dabek [1], Jaroslaw Szrek [2], Radoslaw Zimroz [1] and Jacek Wodecki [1,*]

[1] Department of Mining, Faculty of Geoengineering, Mining and Geology, Wrocław University of Science and Technology, 50-370 Wroclaw, Poland; przemyslaw.dabek@pwr.edu.pl (P.D.); radoslaw.zimroz@pwr.edu.pl (R.Z.)
[2] Department of Fundamentals of Machine Design and Mechatronic Systems, Faculty of Mechanical Engineering, Wroclaw University of Science and Technology, 50-370 Wroclaw, Poland; jaroslaw.szrek@pwr.edu.pl
* Correspondence: jacek.wodecki@pwr.edu.pl

Abstract: Complex mechanical systems used in the mining industry for efficient raw materials extraction require proper maintenance. Especially in a deep underground mine, the regular inspection of machines operating in extremely harsh conditions is challenging, thus, monitoring systems and autonomous inspection robots are becoming more and more popular. In the paper, it is proposed to use a mobile unmanned ground vehicle (UGV) platform equipped with various data acquisition systems for supporting inspection procedures. Although maintenance staff with appropriate experience are able to identify problems almost immediately, due to mentioned harsh conditions such as temperature, humidity, poisonous gas risk, etc., their presence in dangerous areas is limited. Thus, it is recommended to use inspection robots collecting data and appropriate algorithms for their processing. In this paper, the authors propose red-green-blue (RGB) and infrared (IR) image fusion to detect overheated idlers. An original procedure for image processing is proposed, that exploits some characteristic features of conveyors to pre-process the RGB image to minimize non-informative components in the pictures collected by the robot. Then, the authors use this result for IR image processing to improve SNR and finally detect hot spots in IR image. The experiments have been performed on real conveyors operating in industrial conditions.

Keywords: image analysis; hot spot detection; image fusion; inspection robotics; belt conveyor

Citation: Dabek, P.; Szrek, J.; Zimroz, R.; Wodecki, J. An Automatic Procedure for Overheated Idler Detection in Belt Conveyors Using Fusion of Infrared and RGB Images Acquired during UGV Robot Inspection. *Energies* **2022**, *15*, 601. https://doi.org/10.3390/en15020601

Academic Editors: Sheng-Qi Yang, Sergey Zhironkin, Monika Hardygora, Daniela Marasová and Mirosław Bajda

Received: 3 November 2021
Accepted: 5 January 2022
Published: 14 January 2022

Publisher's Note: MDPI stays neutral with regard to jurisdictional claims in published maps and institutional affiliations.

1. Introduction

Belt conveyors are used to transport bulk materials over long distances. Due to the harsh environmental conditions, the elements of the conveyor are subject to accelerated degradation processes and must be monitored. One of the key problems is monitoring the degradation of thousands of idlers installed to support the belt. Dusty conditions, high humidity, impulsive load (when oversized pieces of material on the belt pass the idler), temporal overloading, etc., may cause really accelerated degradation of the coating of the idler and rolling element bearings installed inside the idler in order to give the rolling ability. One of the most popular ways to evaluate the condition of the idler is its visual inspection, temperature measurement, and acoustic sound analysis. In the paper, we propose to use both RGB images and IR images from infrared cameras.

Even though the conveyor seems to be a simple mechanical system, it may happen that other hot elements may appear around an idler. Thus, we propose original procedure for image processing that exploit some characteristic features of conveyors to pre-process the RGB image to minimize non-informative components in the pictures collected by robots. Then, we use this result for IR image processing to improve signal to noise ratio (SNR), and finally, we detect hot spots in the IR images.

In our previous paper [1], we proposed a fusion of information from idler detection in RGB and IR images.

However, we have found that to be applied in real conditions, the method should be significantly extended due to the number of potential hot spots related to other sources. Thus, we proposed a combination of pre-processing of RGB images, processing IR images based on results of pre-processing, and finally, an analysis of IR images in order to detect hot spots related to the idler.

The method proposed in this paper is based on solutions that operate on two different, timewise synchronized datasets: RGB and IR images. The first solution consists of searching of a decreased region of interest (ROI) to minimize the number of potential artifacts, while the second part is focused solely on the thermal detection of damaged idlers. There are many available techniques to filter the image (image smoothing or edge filtering). A probabilistic Hough transform is popular in image analysis applications. It has been proposed here to crop out from the image exact area occupied by belt conveyor. This area is detected taking advantage of the fact that the outer edges of the belt can be simplified to straight lines.

The Hough transform is a commonly used algorithm, applicable both in case of simple shapes (lines, circles, etc.) easily visible in images and more complicated cases where contours become obstructed or take less regular forms. The most popular example in recent years, due to extensive research in the topic of car autonomy, is road lane tracking. It was presented among others in [2], where progressive probabilistic Hough transform helped to detect boundaries, keeping the vehicle on the road, or in [3], where it was discussed in terms of practical use and existing problems. Fruit detection in a natural environment presented in [4] is an example where a complex problem was solved by merging the Hough transform with other methods of shape detection. A solution similar to the one presented in this paper can be found in [5], where Hough transform and color-based detection were used for intelligent wielding flame recognition. In [6], a problem directly related to the topic of this paper was presented. Authors incorporated deep learning methods both in order to crop the belt conveyor and to detect idlers. Only after the initial ROI reduction was Hough transform applied in order to detect the lines that describe the left and right edges of the belt.

Hot spot detection applied on reduced ROI has been performed by a color-based blob detection algorithm. Blobs can be understood as a group of connected pixels that share one or more properties. There are many ways that blobs can be defined and localized [7], a main difference coming from the pixel group properties used in the detection process. In the past years, many improvements to the method were proposed, in terms of color detection performance (such as in the case of [8]) or by adapting modern CNN solutions to blob detection algorithm, as can be seen in agricultural applications [9]. The problem of hot spot detection in the case of belt conveyors was also presented in [10], being a natural extension of an article cited earlier [6]. The method implemented there takes advantage of the idler outer edge shape and temperature difference between the heating idler and its background.

An exemplary use of blob detection can be seen largely in the biomedical field (such as feature detection in medical images discussed in [11] and more narrow applications in the form of breast tumor detection in [12]) or industrial applications, such as automated defect detection in the material structure [13].

Nowadays, it is hard to imagine the existence of modern industries without some kind of assistance from systems operating on visual data. Ongoing research in fields of quality control [14–16]), medicine ([17–19], or agriculture ([20–22]) are just a few examples outside of the mining industry.

Further development in possible applications requires the simultaneous development of implemented tools, such as in the case of this paper, merging information from connected datasets: RGB and infrared images. Research on infrared image data as source of additional information obtainable through analysis was conducted alongside standard visual data analysis, as it can usually benefit from the same methods, although with complications. Solutions to improve quality of processed infrared data are still being researched, such as edge detection supported by spiking neural networks [23] or contrast enhancement based on memristive mapping [24].

In [25], the problem of infrared images segmentation was discussed, with a proposed solution emerging from a combination of infrared images and a depth map computed from two (stereo) images set on different angles. In [26], thermal infrared image analysis was proposed as a method of detecting breathing disorders , while in [27], the thermal inspection of animal welfare was presented with examples based on farm animal footage.

The fusion of infrared and RGB data is another topic going through intensive research. The goal of this operation can be stated as obtaining complementary data with abundant and detailed information in visual images and effectively small areas in IR that allow for the desired detection to be performed [28]. Road safety oriented solutions can be implement as in [29], where pedestrian detection was divided to night and day cycles, using visual camera for the former and infrared camera for the latter. An example of image fusion powered by neural convolutional network and its possibilities was presented in [30].

The identification of the damage in mine infrastructure via image analysis (RGB, IR) was presented in [31]. The authors analyzed the existing idlers inspection techniques and proposed their own solution based on an Unmanned Aerial Vehicle and an RGB and IR camera sensory system. In this case, the detection of objects was based on the ACF method—shallow learning techniques, which gave good results, but there is still room for increasing the effectiveness of operation. Pattern recognition algorithms, segmentation, feature extraction, and classification for the automatic identification of the typical elements on infrared images were used in [32]. The sensory system has been mounted on a mobile platform that moves along the conveyor. The measuring system has been integrated into the system for forecasting damage to the drive system and the transmission of the belt conveyor drive, based on the temperature measurement and recorded operating data, including failures.

Another solution combining the deep learning algorithm and the image processing technology was proposed in [6]. In addition, audio data were used to locate defects [33]. The deep-convolution-network-based algorithms for belt edge detection were presented in [34]. IR sensors are also used to detect leaks by analyzing a disturbance in the temperature profile in the observed infrastructure [35].

IR cameras have been already discussed in many papers. Szurgacz et al. [36] used IR cameras to evaluate the condition of a conveyor drive unit. A similar problem has been discussed in [37]. Gearboxes used in the drive system of the belt conveyor located in the underground mine were tested using the thermal imaging method [37]. In stationary operation, the temperature characteristics of the object were created as a reference for later tests. In [38], using IR cameras, the authors tried to correlate the temperature of conveyor belt (in a pipe conveyor) to the value of the tensioning force. Skoczylas [39] used acoustic data collected by a legged robot to maintain belt conveyors used in a mineral processing plant.

The authors in [40] proposed a system which combines a thermal-infrared camera with a range sensor to obtain three-dimensional (3D) models of environments with both appearance and temperature information. The proposed system creates a map by integrating RGB and IR images through timestamp-based interpolation and information obtained at the system calibration stage. The authors emphasize that the system is able to accurately reproduce the surroundings, even in complete darkness.

Collecting the images by mobile robot may be a significant factor influencing image analysis. It should be mentioned that the idea of replacing of human inspection by inspection robots is not new. Inspection robots are also used in other applications, for example, for the inspection of airports [41]. Raviola [42] discussed using collaborative robots to enhance Electro-Hydraulic Servo-Actuators used in aircrafts. A legged robot was proposed for inspection in [43], a wheeled robot equipped with a manipulator was discussed in [44], and finally, a specially designed robot with a self-leveling system and hybrid wheel-legged locomotion was discussed in [45]. Other solutions based on robots suspended on a rope above the conveyor [46] or UAV [31] are also interesting solutions.

To summarize, there is a strong need for robot-based inspection (instead of human-based), and an automatic procedures for data analysis are required. This is a very important issue from the practical point of view, especially in very harsh environments, such as the mining industry. Thus, in this paper, we proposed a solution for belt conveyor inspection based on UGV. After data acquisition, the data should be processed automatically; thus, a procedure for hot spot detection is proposed. The originality of the method is related to a combination of RGB and IR image pre-processing in order to minimize ROI and the number of potential artifacts.

The structure of the paper is as follows: First, we describe the experiment in the mine and the mobile inspection platform used to acquire the data. Then, we propose an original methodology that results in hot spot detection. Finally, the method has been applied to analyze real data captured during an inspection mission.

2. Problem Formulation

The purpose of the paper is to develop an automatic procedure for overheated idler detection in belt conveyors based on infrared and RGB image fusion. Images have been acquired by UGV inspection robots during field missions. "The Diagnostic Information" is included in the IR image, however, there are many other objects with potentially similar or higher temperatures. Thus, the idea applied here is to combine RGB and IR images. We use RBG images to "filter out" irrelevant data (that means segmentation of the image) before proper hot spot detection. Then, we extracted the ROI to the IR image, and for a limited area of IR we detect hot spots in IR image domain.

3. Experiment Description

The experiments were carried out on a real belt conveyor in an opencast minerals mine, close to the bunker where transported material is stored. A mobile robot equipped with sensor systems was going along the conveyor to record the image from the RGB and IR cameras. The cameras were directed towards the conveyor in order to better cover the area of interest. Then, the data were analyzed in order to detect damaged idlers (overheating).

3.1. The Belt Conveyor

The belt conveyor analyzed in this paper is a mechanical system used for the continuous horizontal transport of raw materials from the mine pit to the bunker. In this case, it was low alumina clays used as aluminosilicate refractory materials. The investigated object was several hundred meters long, and it operates in noisy environment. The design of the conveyor is classical. The key problem was to identify overheated idlers as potential source of fire.

3.2. The UGV Inspection Robot

The platform has four wheels in a skid steering approach. It is driven by two direct current (DC) motors integrated with the gear. The power transmission is performed by a toothed belts, one on each side. During the inspection, the platform was remotely controlled from the user's panel. The measurement system, recording a series of data, was also started remotely. However, a fully autonomous version is the ultimate goal for inspection. The robot is presented in Figure 1.

Visualization of the data recorded by the measurement system is presented in Figure 2. During the inspection, various data were recorded, including spatial, but in this case, we focus on damage detection using only RGB and IR images from the cameras.

Basic parameters of the cameras used during the experiment can be seen in Table 1. It is worth noting that the cameras' frames per second (FPS) parameters varied in the range of 24–26 fps due to robot computer being under heavy information load coming from many different sensors mounted on the platform. No shifts of frames between both cameras have been detected. Value stated in the table is averaged.

Figure 1. View of the robot during inspection.

Figure 2. View in Rviz of recorded data during inspection. (**Top left**): depth camera data (gray) and lidar scan lines (red); (**top right**): front overview; (**bottom left**): depth camera data and lidar scan lines mapped onto 2D frontal view; (**bottom center**): RGB camera frame; (**bottom right**): IR camera frame (flipped horizontally).

Table 1. Camera parameters.

Parameter	RGB and IR Camera
Frames per second	25 fps
Resolution	640×480
Mounting height	100 cm above shelf
Observation angle	$40°$

3.3. The Inspection Mission

An experiment carried out in the mine resulted in the collection of diagnostic data acquired by the UGV inspection robot developed under Autonomous Monitoring and Control System for Mining Plants (AMICOS) project [47]. Various types of data were recorded (RGB images, infrared thermography images, noise, lidar data, etc.) for the purposes of various diagnostic tasks. Figure 3 shows a frame from the captured movie of a conveyor in the hall, the next Figure 4 shows an analogous view but recorded with an infrared camera.

The conditions under which the experiment was conducted were quite difficult from the point of view of image analysis. Observing raw IR images, one may notice many potential hot spots, or in general, regions that could disturb the analysis, such as the warm ceiling of the hall (this is not diagnostic information). Thus, it is clear that pre-processing is needed to clean up the image before the actual hot spot detection procedure.

Basic parameters of the belt conveyor can be seen in Table 2. It is worth noting that inspection had not started at the very beginning of the belt conveyor.

Table 2. Belt conveyor parameters.

Parameter	Value
Conveyor length	150 m
Belt width	800 mm
Idler diameter	133 mm
Idler spacing	1.45 m

Figure 3. Example of RGB scene.

Figure 4. Example of IR scene.

4. Methodology for Hot Spot Detection

To achieve an efficient diagnostic procedure for overheated idler detection, a particular approach of RGB and IR image collection, pre-processing, and analysis is proposed. The approach consists of several consecutive steps, beginning with the image area reduction by cutting the upper half of the image. This decision is based on the camera angle applied during the robot ride and is useful by the means of repeatable empty area removal (in this example, containing windowed walls and ceilings). Of course, for other missions with other objects and different angles, this segmentation step needs to be adjusted accordingly. In this case, the parameter of cropping out the upper segment of the image has been set to the half of the image height. Reducing the size of the ROI allows improving the efficiency of the procedure (elimination of false-positives in unrealistic areas) as well as computational efficiency (smaller region to search). However, in the current implementation, the efficiency of computation does not matter a lot because the procedure is executed offline.

The initial step of pre-processing focuses on extracting the area contour of the belt conveyor from the RGB image. It can be divided into following subgroups (as shown in the Figure 5 flowchart):

- general straight line detection;
- incorrect line removal and line grouping;
- belt conveyor line selection;
- mask generation (based on selected line).

The final product of this process takes the form of a mask that can filter out the background of image, leaving only the conveyor belt for further analysis. Then, the standard color-based heat detection algorithm can be applied for overheating idler detection by using the second half of the dataset: infrared images. Thanks to timewise synchronization and only slight angle difference between used cameras, the mask extracted in the previous part of the method can be translated directly onto the IR image.

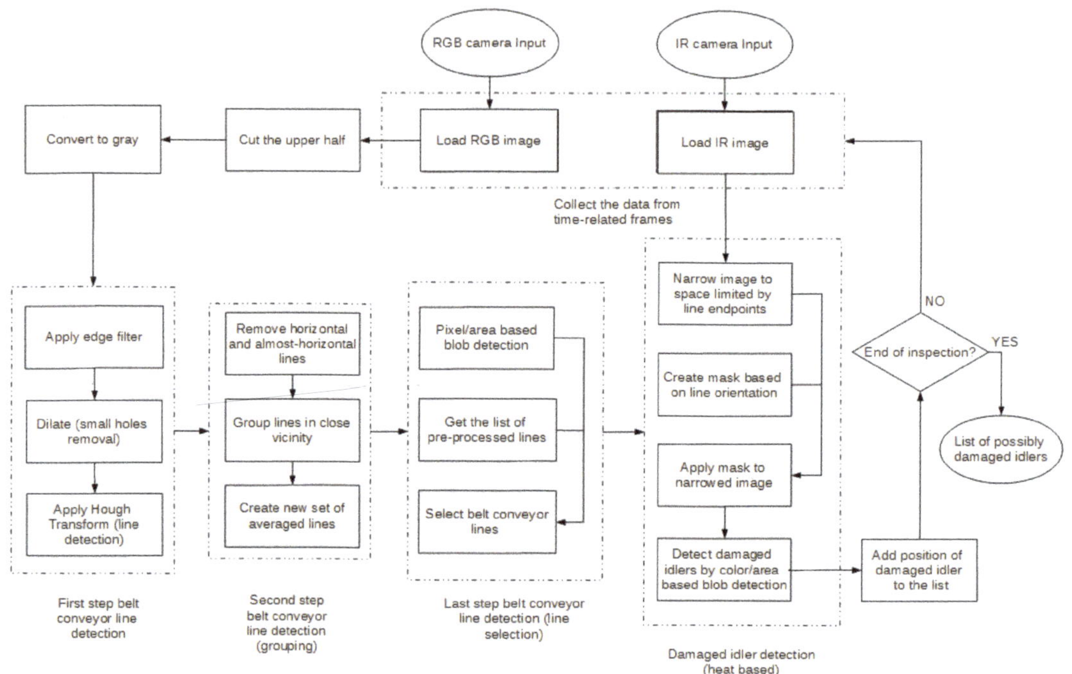

Figure 5. Simplified flowchart of algorithm.

4.1. General Line Detection

For detection purposes, conveyor belt edges can be treated as a straight lines. Two methods of line detection were taken into consideration: Hough transform (HT) and probabilistic Hough transform, also referred to as randomized Hough transform (RHT).

HT can be explained as a voting process where each point belonging to the patterns votes for all the possible patterns passing through that point. These votes are accumulated in accumulator arrays called bins, and the pattern receiving the maximum votes is recognized as the desired pattern [48]. In Hough transform, lines are described by the polar form of line Equation (1), where ρ represents the perpendicular distance of the line from the origin measured in pixels, θ represents the angle of deviation from the horizontal level measured in radians, and (x, y) represent the coordinates of the point that the line is going through:

$$\rho = x\cos(\theta) + y\cos(\theta) \tag{1}$$

Possible lines are deduced by locating local maxima in accumulated parameter space, the size of which depends on the number of pixels inside of an image. Since non-randomized Hough transform algorithms must compute through each pixel separately, the problem becomes one memory and time consumption. In HT, this problem is solved by introduction of random pixel sampling. It allows for significantly increased computation speed, sacrificing some accuracy in the process, depending on the complexity of the required lines and provided data.

In case of data used in this paper, results obtained from both HT and RHT did not show significant differences in terms of accuracy, therefore, the RHT algorithm was chosen, as lower computational time is extremely important in the viewpoint of the mobile platform and its hardware capabilities.

For the RHT algorithm to work correctly, several steps had to be taken before its implementation. Firstly, images were converted to gray scale and smoothed by Gaussian filter in order to remove part of the noise related to the used equipment and the quality of registered data. Then, a Canny filter was applied, as it gave better results than other filters in regard of lines continuity and noise influence. Similar conclusions in terms of filter performance were drawn in [49,50]. An example of an applied Canny filter is shown in Figure 6. The image can be dilated afterwards, as increased line continuity allows for better results from Hough transform.

Figure 6. Exemplary effect of applied Canny filter.

Parameters of detection in this transform are kept in a wide range, as it allows for smoother positioning and grouping in later steps. This especially considers the acceptance of long gaps during line detection and shifting the focus to possibly long line searches. Results of the implemented RHT are presented in Figure 7.

Figure 7. Exemplary effect of general line detection (Hough transform).

4.2. Detected Lines Preliminary Selection and Grouping

During the inspection, the robot was moving along belt conveyor. The camera observation angle was considered as constant. Edge detection approach may provide many lines detected in the picture. Some of them are artifacts, others are not important for the next steps. Each group of lines obtained from the Hough transform can be therefore represented by the single line averaged based on group of lines with similar slope coefficient and position.

In case of the collected data, horizontal and nearly horizontal lines were removed, as they could not describe belt conveyor due to the camera angle. The next step consisted of grouping and merging of the lines in close vicinity (based on the m and y_0 parameter of a general line equation). It should be noted that vertical lines should be also removed (if detected). Results of the preliminary selection and grouping can be seen in Figure 8.

Figure 8. Exemplary effect after line grouping.

4.3. Line Selection Supported by Blob Detection Algorithm

As can be seen in Figure 9, the blob detection algorithm can provide some incorrect detection results, so final idler detection should be supported by additional criteria.

Figure 9. Example of incorrect idler blob detection.

Belt conveyor design contains several longitudinal elements (as the edge of a frame of the conveyor belt). Along these elements, we expect to find hot spots (overheated idler). The relation between the belt edge and the location of idlers may be used in two ways. First, the highest number of detected blobs located close to lines pre-selected in Section 4.2 allows concluding that one of these lines is the belt edge. Second, the distance between the identified belt edge and center of particular blob may be a criterion to remove incorrect blob detection result. Therefore, a combination of blob detection and the distance between blob center and the border line associated with the belt may be a basis to improve hot spot detection efficiency.

4.4. Mask Creation and Implementation

The first step to reduce ROI focuses on cropping out from IR image the area described by two corners. These corners are based on coordinates of selected line endpoints. Since the IR and RGB dataset were recorded with similar angles and synchronized timewise, the line endpoint positions can be directly translated between images, with a possibility of small but acceptable mismatch.

After that, a mask is created, with the initial ROI based on the previously cropped area. Borders of the mask are defined either by edge lines detected in previous process (as shown in Figure 8) or by the use of a predetermined shape. The predetermined shape takes the central diagonal (corner to corner) line as a base and creates constant borders based on the regular shape of belt that becomes smaller the further it is from the robot. An example of the created mask can be seen in Figure 10.

Figure 10. Exemplary mask created in the process of ROI reduction.

Then, the mask is applied on the IR image, creating an even smaller ROI. It limits the risk of false hot spot detection coming from the background, such as either working machines or materials prone to heating from different sources (heat transfer or exposure to sunlight). The result of a narrowed frame with the applied mask is presented in Figure 11.

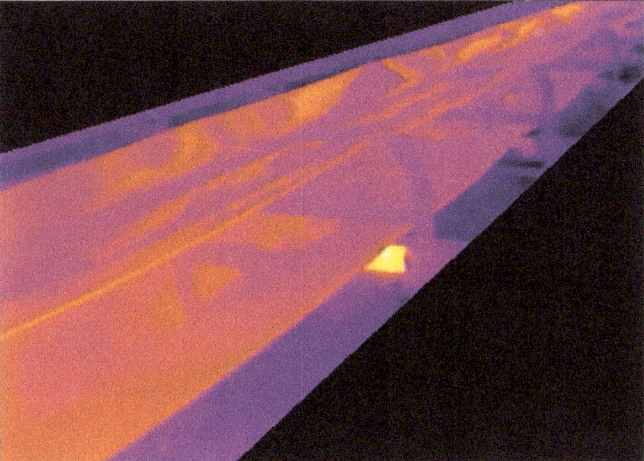

Figure 11. Frame narrowed to the main line with applied mask.

4.5. Heat-Based Damage Detection

The last step of the method focuses directly on hot spot detection. Thanks to ROI reduction, the possibility of incorrect classification is reduced, and simple methods can be implemented with higher rates of successful detection. This method uses the blob detection technique. In opposite to the previously implemented one (in the blob supported line selection), this time it is based on area and color parameters.

Main idler issues present in form of generated heat that can occur during inspection come from the following:

- heat generated during high friction contact between idler and belt;
- damaged idler bearing that causes the idler to stop;

- partially damaged bearing of the idler that causes the core of the idler to become hot without stopping the idler rotation.

The solution to find the potentially damaged idlers is based on combination of:

- focus on restricted color range;
- small, circular area of counted pixels;
- assumption that no other source of heat than the idler/belt friction can be found in the previously processed ROI.

To distinguish areas that can be recognized as damaged idlers, a color-based threshold was used. This process takes place on the hue-saturation-value (HSV) converted image. Hue, a parameter responsible for the color differentiation, was set to the bright orange to bright yellow range with the rest of the parameters set to a possibly wide range of intensities.

The thresholded image was then passed to the blob detection algorithm, focused on area detection wide enough to detect mid- to close-range passing idlers, but narrow enough to not consider bigger blobs created by heated belt.

Examples of final results in the form of detected idlers can be seen in Figures 12–15.

As can be seen in Figures 13 and 15, some potentially damaged idlers are not detected over a single frame. It is due to the distance and fact that too small hot spots cannot be considered in process of detection, since noise (such as heating belt) can create spots with similar properties. It can be solved by reducing the ROI to lower x-axis values.

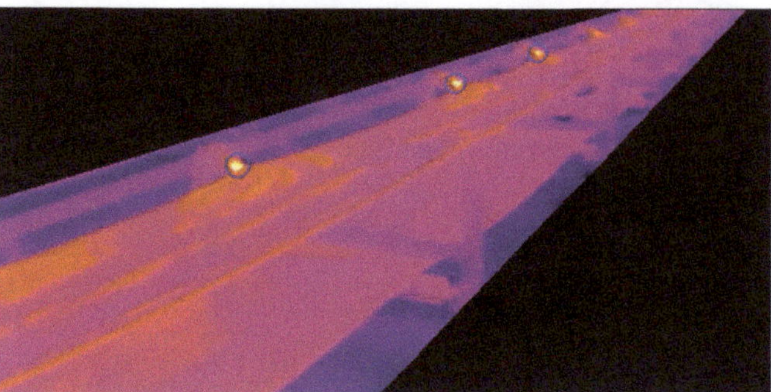

Figure 12. Result of damaged idler detection, example 1.

Figure 13. Result of damaged idler detection, example 2.

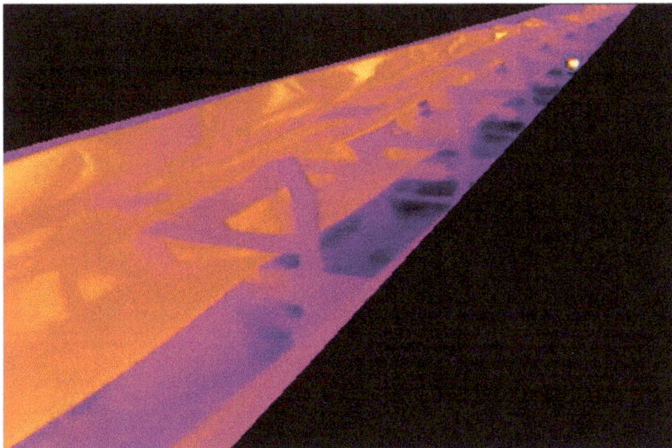

Figure 14. Result of damaged idler detection, example 3.

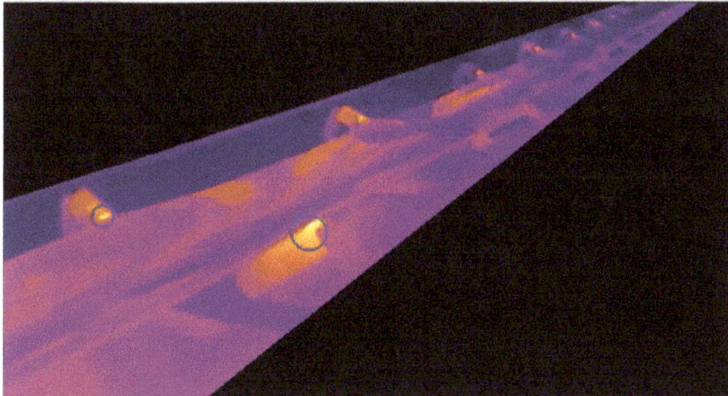

Figure 15. Result of damaged idler detection, example 4.

For better visualization of the method results, cropped ROI with detected idlers was applied on the original RGB frame in Figure 16. The differently colored part of the shown image represents an actual ROI cropped out by the proposed algorithm.

Figure 16. Algorithm result visualisation on starting data.

As can be seen, angle difference between the RGB and IR cameras resulted in slightly shifted perspective that had to be taken into consideration during boundary creation. In case of not detected or incorrectly detected belt conveyor edges, the predetermined shape of the mask was used as substitution. Later steps do not deviate from the described method. The predetermined shape of the mask can be seen in Figure 17 and example result in Figure 18.

Figure 17. Predetermined mask shape.

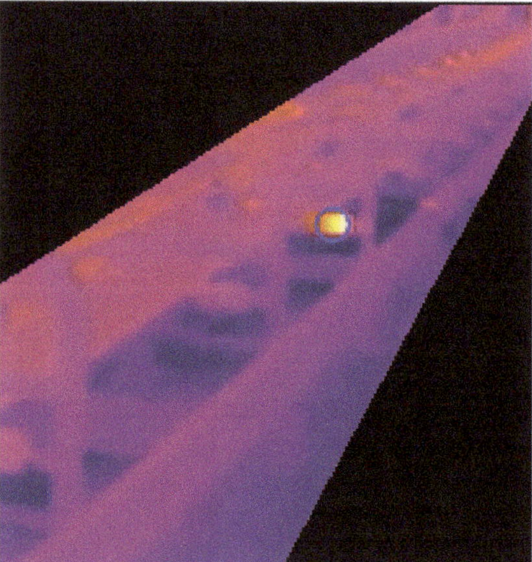

Figure 18. Possible effect of applied predetermined mask shape.

The ratio between the frames with correctly detected lines and overall computed frames has been summarized in Table 3. Correct line detection represent overall number of instances where the line detection worked properly and no predetermined shape had to be applied. Analogically predetermined shape cases represent the summarized number of instances where the line detection part of the algorithm failed to obtain lines that could correctly describe belt conveyor.

Table 3. Line detection results.

Result	Value
Computed frames	1800
Correct line detection	1699
Predetermined shape cases	101
Correct detection ratio	94.39%

5. Results

The method presented in this paper was tested on a dataset consisting of several videos. Each detection case consisted of 1500+ frames with at least 20 different idlers present during the robot ride. Each detection case included both faulty and healthy idlers. Some sections of the belt conveyor did not have faulty idlers, which, in terms of the dataset, can be understood as some frames contained only healthy idlers. Therefore, susceptibility to false detections in these cases was also included in the research process. All frames were taken during the simulated robot inspection, where the robot was constantly moving along the belt conveyor.

The results described in this section focus on the video that presented the most difficulties to the algorithm. The trend of at least 10 frames late detection that occurred during all tested datasets can be explained by the parameters that were used in hot spot detection algorithm. These parameters were set to detect only objects of defined size, as a way to prevent false detection related to noise coming through the process.

The main problems of the algorithm performance of hot spot detection resulted from the increased influence of belt temperature. The algorithm was especially inefficient in correct classification in cases where high temperatures persisted over long fragments of the belt conveyor. In those cases, the number of invalid hits and missed idlers raised significantly.

In order to "measure" the quality of faulty idler detection, we use classical parameters in pattern recognition theory: sensitivity and specificity.

Sensitivity (TPR—True Positive Rate) was defined in (2), where true-positive (TP) represents the number of true positive cases and false-negative (FN) represents the number of false negative cases:

$$\text{TPR} = \frac{\text{TP}}{\text{TP} + \text{FN}} \tag{2}$$

In terms of this paper, TP can be described as the number of frames where damaged idlers were detected correctly, whereas FN can be described as the number of frames where overestimation has occurred (spot on the image without damaged idler classified as one that contains it). Therefore, sensitivity relates to the algorithm's ability to correctly detect the damaged idlers. The sensitivity of detection performed on the selected part of the dataset totaled at 0.83 after 1800 computed frames.

Specificity (TNR—True Negative Rate) was defined in (3), where TN represents the number of true negative cases and FP represents the number of false positive cases:

$$\text{TNR} = \frac{\text{TN}}{\text{TN} + \text{FP}} \tag{3}$$

In terms of this paper, TN can be described as the number of frames where no damaged idlers were neither present nor detected, whereas FP can be described as the number of frames where underestimation has occurred (damaged idler was not detected by the algorithm). Therefore, the specificity relates to the algorithm's ability to correctly reject spots in the computed images that do not contain damaged idlers. The sensitivity of the detection performed on the selected dataset totaled at 0.74 after 1800 computed frames.

Accuracy (ACC) was defined in (4), where TP and TN represent the same parts of the data that were described in sensitivity and specificity, whereas N represents the total number of computed frames:

$$\mathrm{ACC} = \frac{\mathrm{TP} + \mathrm{TN}}{\mathrm{N}} \tag{4}$$

Accuracy can be therefore described as the ratio between the number of correct assessments and the number of computed frames. The accuracy of the detection performed on the selected dataset totaled at 0.78 after 1800 computed frames.

The summarized results are presented in Figure 19, as accuracy in correlation to computed frames. For better understanding, data were provide in the form of a table with absolute numbers of detection cases in Table 4.

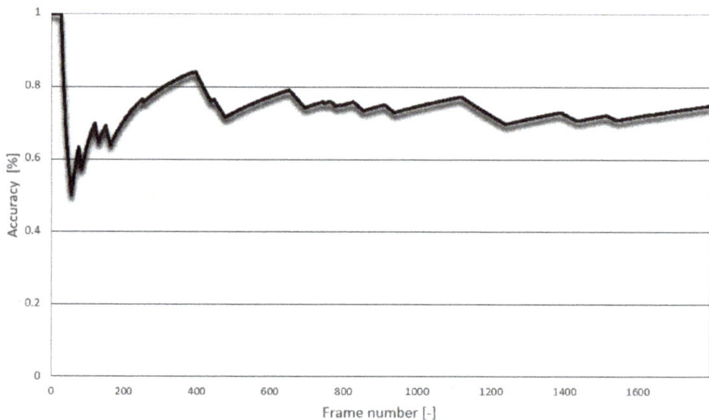

Figure 19. Detection accuracy timeline during inspection.

Table 4. Accuracy parameters.

Number of Frames	TP	TN	FN	FP
1800	688	659	135	36

As can be seen in Figure 19, the main reason for lowered detection efficiency occurred mostly in the first part of the analyzed video. As stated in the second paragraph of this section, the case of high belt temperature over a long period of time was present. The exemplary frame can be seen in Figure 20.

The difference between this situation and similar instances present through the rest of the video comes from the temperature of idlers that remained indistinguishable from the temperature of the belt surface. It resulted in an increased number of both false detections and potentially missed damaged idlers. An example of a case where detection works properly despite the increased belt temperature can be seen in Figure 21.

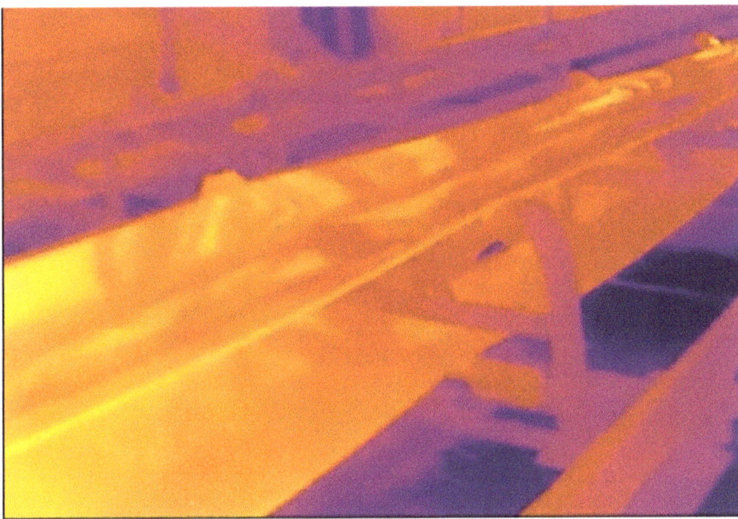

Figure 20. Exemplary frame with problematic case.

Figure 21. Exemplary frame with problem solvable by algorithm.

6. Conclusions

A procedure for image processing and information fusion based on datasets acquired during belt conveyor inspection was proposed in this paper. Information was collected by sensors mounted on a mobile robot: RGB and IR cameras, among others. The reason for replacing human by autonomous robots has been highlighted in many papers—dangerous environment, noisy, dusty atmosphere, long distances, number of elements to asses, etc. Robots can do it automatically with established criteria and instant database completion. The methodology uses a combination of specific, distinctive features detectable on belt conveyors, techniques widely used in image processing, and their adaptation to the task of detecting overheated spots.

The measurements have been conducted in real conditions present in a mine during the robot ride taken along the belt conveyor and selected from different rides taken at different times. The first part of the proposed algorithm for RGB data analysis focused on cropping the belt conveyor out of the image and gave positive results in terms of ROI reduction. The second part of the proposed algorithm focused on direct hot spot detection. The result of RGB image cropping has been used for ROI identification in IR images. Then, the blob-based hot spot detection algorithm was applied. The results of hot spot detection proved to be sufficient for proper damaged idler classification in most of tested cases. The results presented in the paper are representative for the worst case scenario selected from the available dataset. With this approach, it is easier to evaluate difficult environmental impacts on the presented method performance.

Future work assumes the improvement of the algorithm in terms of adaptability to bigger changes in environment temperature. Since all measurements were conducted in same season, it is safe to presume that the described algorithm would give different results in, for example, colder parts of the year. Solutions to that can be looked for in adaptable recognition of temperature level. It can be based on the detected difference between the temperature of working elements and elements of background noise. In this paper, the hot spot detection part of the algorithm is based solely on two-step recognition, taking the color (without considering the color of the background) and number of pixels in close proximity.

Author Contributions: Conceptualization, P.D., J.S. and R.Z.; methodology, P.D. and J.S.; software, P.D.; validation, J.S., P.D., J.W. and R.Z.; formal analysis, J.S.; investigation, P.D. and J.S.; resources, R.Z.; data curation, J.S.; writing—original draft preparation, J.S., J.W. and P.D.; writing—review and editing, J.W. and R.Z.; visualization, J.S.; supervision, R.Z.; project administration, R.Z.; funding acquisition, R.Z. All authors have read and agreed to the published version of the manuscript.

Funding: This activity has received funding from the European Institute of Innovation and Technology (EIT), a body of the European Union, under the Horizon 2020, the EU Framework Programme for Research and Innovation. This work is supported by EIT RawMaterials GmbH under Framework Partnership Agreement No. 19018 (AMICOS. Autonomous Monitoring and Control System for Mining Plants).

Institutional Review Board Statement: Not applicable.

Informed Consent Statement: Not applicable.

Data Availability Statement: Archived data sets cannot be accessed publicly according to the NDA agreement signed by the authors.

Acknowledgments: Supported by the Foundation for Polish Science (FNP).

Conflicts of Interest: The authors declare no conflict of interest.

References

1. Szrek, J.; Wodecki, J.; Błazej, R.; Zimroz, R. An inspection robot for belt conveyor maintenance in underground mine-infrared thermography for overheated idlers detection. *Appl. Sci.* **2020**, *10*, 4984. [CrossRef]
2. Marzougui, M.; Alasiry, A.; Kortli, Y.; Baili, J. A Lane Tracking Method Based on Progressive Probabilistic Hough Transform. *IEEE Access* **2020**, *8*, 84893–84905. [CrossRef]
3. Huang, Q.; Liu, J. Practical limitations of lane detection algorithm based on Hough transform in challenging scenarios. *Int. J. Adv. Robot. Syst.* **2021**, *18*, 17298814211008752. [CrossRef]
4. Lin, G.; Tang, Y.; Zou, X.; Cheng, J.; Xiong, J. Fruit detection in natural environment using partial shape matching and probabilistic Hough transform. *Precis. Agric.* **2020**, *21*, 160–177. [CrossRef]
5. Chen, W.; Chen, S.; Guo, H.; Ni, X. Welding flame detection based on color recognition and progressive probabilistic Hough transform. *Concurr. Comput. Pract. Exp.* **2020**, *32*, e5815. [CrossRef]
6. Liu, Y.; Miao, C.; Li, X.; Xu, G. Research on Deviation Detection of Belt Conveyor Based on Inspection Robot and Deep Learning. *Complexity* **2021**, *2021*, 3734560. [CrossRef]
7. Kaspers, A. Blob Detection. Master's Thesis, Utrecht University, Utrecht, The Netherlands, 2011.
8. Khanina, N.; Semeikina, E.; Yurin, D. Scale-space color blob and ridge detection. *Pattern Recognit. Image Anal.* **2012**, *22*, 221–227. [CrossRef]
9. Chen, S.W.; Shivakumar, S.S.; Dcunha, S.; Das, J.; Okon, E.; Qu, C.; Taylor, C.J.; Kumar, V. Counting Apples and Oranges with Deep Learning: A Data-Driven Approach. *IEEE Robot. Autom. Lett.* **2017**, *2*, 781–788. [CrossRef]
10. Liu, Y.; Miao, C.; Li, X.; Ji, J.; Meng, D. Research on the fault analysis method of belt conveyor idlers based on sound and thermal infrared image features. *Measurement* **2021**, *186*, 110177. [CrossRef]
11. Han, K.T.M.; Uyyanonvara, B. A Survey of Blob Detection Algorithms for Biomedical Images. In Proceedings of the 2016 7th International Conference of Information and Communication Technology for Embedded Systems (IC-ICTES), Bangkok, Thailand, 20–22 March 2016; pp. 57–60.
12. Moon, W.K.; Shen, Y.W.; Bae, M.S.; Huang, C.S.; Chen, J.H.; Chang, R.F. Computer-aided tumor detection based on multi-scale blob detection algorithm in automated breast ultrasound images. *IEEE Trans. Med. Imaging* **2012**, *32*, 1191–1200. [CrossRef] [PubMed]
13. Czimmermann, T.; Ciuti, G.; Milazzo, M.; Chiurazzi, M.; Roccella, S.; Oddo, C.M.; Dario, P. Visual-based defect detection and classification approaches for industrial applications—A survey. *Sensors* **2020**, *20*, 1459. [CrossRef] [PubMed]

14. Muniategui, A.; de la Yedra, A.G.; del Barrio, J.A.; Masenlle, M.; Angulo, X.; Moreno, R. Mass production quality control of welds based on image processing and deep learning in safety components industry. In Proceedings of the Fourteenth International Conference on Quality Control by Artificial Vision, Mulhouse, France, 15–17 May 2019; Cudel, C., Bazeille, S., Verrier, N., Eds.; International Society for Optics and Photonics: Bellingham, WA, USA, 2019; Volume 11172, pp. 148–155.

15. Liu, C.; Law, A.C.C.; Roberson, D.; Kong, Z.J. Image analysis-based closed loop quality control for additive manufacturing with fused filament fabrication. *J. Manuf. Syst.* **2019**, *51*, 75–86. [CrossRef]

16. Wojnar, L. *Image Analysis: Applications in Materials Engineering*; CRC Press: Boca Raton, FL, USA 2019; pp. 213–216.

17. Wang, F.; Casalino, L.P.; Khullar, D. Deep learning in medicine—Promise, progress, and challenges. *JAMA Intern. Med.* **2019**, *179*, 293–294. [CrossRef] [PubMed]

18. Poostchi, M.; Silamut, K.; Maude, R.J.; Jaeger, S.; Thoma, G. Image analysis and machine learning for detecting malaria. *Transl. Res.* **2018**, *194*, 36–55. [CrossRef] [PubMed]

19. Choi, H. Deep learning in nuclear medicine and molecular imaging: Current perspectives and future directions. *Nucl. Med. Mol. Imaging* **2018**, *52*, 109–118. [CrossRef] [PubMed]

20. Chiu, M.T.; Xu, X.; Wei, Y.; Huang, Z.; Schwing, A.G.; Brunner, R.; Khachatrian, H.; Karapetyan, H.; Dozier, I.; Rose, G.; et al. Agriculture-vision: A large aerial image database for agricultural pattern analysis. In Proceedings of the IEEE/CVF Conference on Computer Vision and Pattern Recognition, Seattle, WA, USA, 13–19 June 2020; pp. 2828–2838.

21. Tsouros, D.C.; Bibi, S.; Sarigiannidis, P.G. A review on UAV-based applications for precision agriculture. *Information* **2019**, *10*, 349. [CrossRef]

22. Sharif, M.; Khan, M.A.; Iqbal, Z.; Azam, M.F.; Lali, M.I.U.; Javed, M.Y. Detection and classification of citrus diseases in agriculture based on optimized weighted segmentation and feature selection. *Comput. Electron. Agric.* **2018**, *150*, 220–234. [CrossRef]

23. Wang, B.; Chen, L.; Zhang, Z. A novel method on the edge detection of infrared image. *Optik* **2019**, *180*, 610–614. [CrossRef]

24. Wang, B.; Chen, L.; Liu, Y. New results on contrast enhancement for infrared images. *Optik* **2019**, *178*, 1264–1269. [CrossRef]

25. Kluwe, B.; Christian, D.; Miknis, M.; Plassmann, P.; Jones, C. Segmentation of Infrared Images Using Stereophotogrammetry. In *VipIMAGE 2017*; Tavares, J.M.R., Natal Jorge, R., Eds.; Springer International Publishing: Cham, Switzerland, 2018; pp. 1025–1034.

26. Procházka, A.; Charvátová, H.; Vyšata, O.; Kopal, J.; Chambers, J. Breathing analysis using thermal and depth imaging camera video records. *Sensors* **2017**, *17*, 1408. [CrossRef]

27. Nääs, I.A.; Garcia, R.G.; Caldara, F.R. Infrared thermal image for assessing animal health and welfare. *J. Anim. Behav. Biometeorol.* **2020**, *2*, 66–72. [CrossRef]

28. Zhao, C.; Huang, Y.; Qiu, S. Infrared and visible image fusion algorithm based on saliency detection and adaptive double-channel spiking cortical model. *Infrared Phys. Technol.* **2019**, *102*, 102976. [CrossRef]

29. González, A.; Fang, Z.; Socarras, Y.; Serrat, J.; Vázquez, D.; Xu, J.; López, A.M. Pedestrian detection at day/night time with visible and FIR cameras: A comparison. *Sensors* **2016**, *16*, 820. [CrossRef] [PubMed]

30. Liu, Y.; Miao, C.; Ji, J.; Li, X. MMF: A Multi-scale MobileNet based fusion method for infrared and visible image. *Infrared Phys. Technol.* **2021**, *119*, 103894. [CrossRef]

31. Carvalho, R.; Nascimento, R.; D'Angelo, T.; Delabrida, S.; Bianchi, A.G.C.; Oliveira, R.A.R.; Azpúrua, H.; Uzeda Garcia, L.G. A UAV-Based Framework for Semi-Automated Thermographic Inspection of Belt Conveyors in the Mining Industry. *Sensors* **2020**, *20*, 2243. [CrossRef] [PubMed]

32. Yang, W.; Zhang, X.; Ma, H. An inspection robot using infrared thermography for belt conveyor. In Proceedings of the 2016 13th International Conference on Ubiquitous Robots and Ambient Intelligence (URAI), Xian, China, 19–22 August 2016; pp. 400–404.

33. Kim, S.; An, D.; Choi, J.H. Diagnostics 101: A Tutorial for Fault Diagnostics of Rolling Element Bearing Using Envelope Analysis in MATLAB. *Appl. Sci.* **2020**, *10*, 7302. [CrossRef]

34. Liu, Y.; Wang, Y.; Zeng, C.; Zhang, W.; Li, J. *Edge Detection for Conveyor Belt Based on the Deep Convolutional Network*; Jia, Y., Du, J., Zhang, W., Eds.; Springer: Singapore, 2019; pp. 275–283.

35. Kroll, A.; Baetz, W.; Peretzki, D. On autonomous detection of pressured air and gas leaks using passive IR-thermography for mobile robot application. In Proceedings of the 2009 IEEE International Conference on Robotics and Automation, Kobe, Japan, 12–17 May 2009; pp. 921–926.

36. Szurgacz, D.; Zhironkin, S.; Vöth, S.; Pokorný, J.; Sam Spearing, A.; Cehlár, M.; Stempniak, M.; Sobik, L. Thermal imaging study to determine the operational condition of a conveyor belt drive system structure. *Energies* **2021**, *14*, 3258. [CrossRef]

37. Błazej, R.; Sawicki, M.; Kirjanów, A.; Kozłowski, T.; Konieczna, M. Automatic analysis of thermograms as a means for estimating technical of a gear system. *Diagnostyka* **2016**, *17*, 43–48.

38. Michalik, P.; Zajac, J. Use of thermovision for monitoring temperature conveyor belt of pipe conveyor. *Appl. Mech. Mater.* **2014**, *683*, 238–242. [CrossRef]

39. Skoczylas, A.; Stefaniak, P.; Anufriiev, S.; Jachnik, B. Belt conveyors rollers diagnostics based on acoustic signal collected using autonomous legged inspection robot. *Appl. Sci.* **2021**, *11*, 2299. [CrossRef]

40. Vidas, S.; Moghadam, P.; Bosse, M. 3D Thermal Mapping of Building Interiors using an RGB-D and Thermal Camera. In Proceedings of the 2013 IEEE International Conference on Robotics and Automation, Karlsruhe, Germany, 6–10 May 2013; pp. 2303–2310.

41. Gui, Z.; Zhong, X.; Wang, Y.; Xiao, T.; Deng, Y.; Yang, H.; Yang, R. A cloud-edge-terminal-based robotic system for airport runway inspection. *Ind. Robot.* **2021**, *48*, 846–855. [CrossRef]

42. Raviola, A.; Antonacci, M.; Marino, F.; Jacazio, G.; Sorli, M.; Wende, G. Collaborative robotics: Enhance maintenance procedures on primary flight control servo-actuators. *Appl. Sci.* **2021**, *11*, 4929. [CrossRef]
43. Ramezani, M.; Brandao, M.; Casseau, B.; Havoutis, I. Fallon, M. Legged Robots for Autonomous Inspection and Monitoring of Offshore Assets. In Proceedings of the OTC Offshore Technology Conference, Houston, TX, USA, 4–7 May 2020.
44. Rocha, F.; Garcia, G.; Pereira, R.F.S.; Faria, H.D.; Silva, T.H.; Andrade, R.H.R.; Barbosa, E.S.; Almeida, A.; Cruz, E.; Andrade, W.; et al. ROSI: A Robotic System for Harsh Outdoor Industrial Inspection—System Design and Applications. *J. Intell. Robot. Syst.* **2021**, *103*, 30. [CrossRef]
45. Bałchanowski, J. Modelling and simulation studies on the mobile robot with self-leveling chassis. *J. Theor. Appl. Mech.* **2016**, *54*, 149–161. [CrossRef]
46. Cao, X.; Zhang, X.; Zhou, Z.; Fei, J.; Zhang, G.; Jiang, W. Research on the Monitoring System of Belt Conveyor Based on Suspension Inspection Robot. In Proceedings of the 2018 IEEE International Conference on Real-Time Computing and Robotics (RCAR), Kandima, Maldives, 1–5 August 2018; pp. 657–661.
47. Szrek, J.; Zimroz, R.; Wodecki, J.; Michalak, A.; Góralczyk, M.; Worsa-Kozak, M. Application of the infrared thermography and unmanned ground vehicle for rescue action support in underground mine—The amicos project. *Remote Sens.* **2021**, *13*, 69. [CrossRef]
48. Duan, D.; Xie, M.; Mo, Q.; Han, Z.; Wan, Y. An improved Hough transform for line detection. In Proceedings of the 2010 International Conference on Computer Application and System Modeling (ICCASM 2010), Taiyuan, China, 22–24 October 2010; Volume 2, pp. 354–357.
49. Shrivakshan, G.; Chandrasekar, C. A comparison of various edge detection techniques used in image processing. *Int. J. Comput. Sci. Issues (IJCSI)* **2012**, *9*, 269.
50. Roushdy, M. Comparative study of edge detection algorithms applying on the grayscale noisy image using morphological filter. *GVIP J.* **2006**, *6*, 17–23.

Article

Dynamic Model of Impact Energy Absorption by a Conveyor Belt in Interaction with the Support System

Daniela Marasova [1,*], Miriam Andrejiova [2] and Anna Grincova [3]

[1] Faculty of Mining, Ecology, Process Control and Geotechnology, Technical University of Kosice, Park Komenskeho 14, 042 00 Kosice, Slovakia

[2] Faculty of Mechanical Engineering, Technical University of Kosice, Letna 9, 042 00 Kosice, Slovakia; miriam.andrejiova@tuke.sk

[3] Faculty of Electrical Engineering and Informatics, Technical University of Kosice, Letna 9, 042 00 Kosice, Slovakia; anna.grincova@tuke.sk

* Correspondence: daniela.marasova@tuke.sk; Tel.: +421-55-602-3123

Abstract: Measurements of the dynamic load of conveyor belts of identical strengths were used to evaluate and compare the data for belts with and without a support system. The goal was to identify the effects of the support system in terms of a relative amount of impact energy absorbed by a conveyor belt. A dynamic model was designed based on selected parameters of the impact process. Damage to conveyor belts, caused by the absorption of impact energy, was evaluated using the applied methods of mathematical statistics.

Keywords: rubber-textile conveyor belt; support system; puncture resistance; absorbed energy

Citation: Marasova, D.; Andrejiova, M.; Grincova, A. Dynamic Model of Impact Energy Absorption by a Conveyor Belt in Interaction with the Support System. *Energies* **2022**, *15*, 64. https://doi.org/10.3390/en15010064

Academic Editor: Davide Astolfi

Received: 6 December 2021
Accepted: 21 December 2021
Published: 22 December 2021

Publisher's Note: MDPI stays neutral with regard to jurisdictional claims in published maps and institutional affiliations.

1. Introduction

Due to the fact that conveyor belts are permanently exposed to dynamic load during their utilisation, it is necessary to understand their dynamic characteristics. Dynamic characteristics of conveyor belts affect smooth conveyance, especially in continuous conveyance. Failures caused by conveyor belt malfunctions that result from poor dynamic characteristics consequently incur high financial losses due to conveyor belt ruptures at belt joints or due to damage to conveyor belts [1,2]. Conveyor belts are the most expensive structural elements of belt conveyors and the costs of conveyor belts represent around 40–60% of the total operating costs of belt conveyance of materials [3].

The kinematic and dynamic processes that run at transfer chutes are very complex. As much as 60% of damage to conveyor belts occurs particularly at chutes. Reliability of chutes in belt conveyance represents a serious problem; therefore, there are efforts aimed at finding methods of how to reduce the number of chutes as the sources of potential failures [4]. Hence, conveyance systems with a large number of chutes exhibit higher complexity and are associated with higher energy consumption and maintenance cost. Provided the chutes are used in line with respective operating requirements, their optimal design significantly affects the service life of conveyor belts and associated energy consumption [5]. Chutes are the points where the energy loss occurs due to the dynamic impact and a consequent absorption of impact energy by a conveyor belt [6] and due to rotating pieces of a conveyed material.

Dynamic characteristics of belts existing in a dynamic impact process are difficult to examine during the belt operation; conveyor belts should therefore be examined in laboratory conditions (by simulations) [7,8]. The absorption ability of composite materials is an important parameter that affects dynamic behaviour of the structures [9]. A standard laboratory method for measuring dynamic properties is, for example, measuring the vibration-damping properties of composite materials, including conveyor belts [10]. Various testing devices have been constructed to facilitate laboratory testing of the dynamic impact process; for example, the test equipment at the Technical University of Wroclaw [11]

---

or the test equipment at the Technical University of Košice, which uses a laser distance sensor to measure a trajectory of the falling load [12]. Laboratory research into punctures in steel-cord conveyor belts was carried out by many others, including authors of papers [13] and [14], who have developed a magnetic high-resolution diagnostic device for steel cord conveyor belts. In addition to monitoring belts during their use in belt conveyors, this device can also be used in laboratories for testing puncture resistance of belts.

A support system is an important structural element of the chutes as it eliminates damage to conveyor belts in terms of punctures. At the chutes, material particles change their direction and velocity; this causes wearing of conveyor belts and punctures made by larger pieces of materials [15]. Experimental investigations conducted by many authors indicated that a support system design has a strong effect on energy consumption during the operation of belt conveyors and on damage to conveyor belts [16–20]. Ambriško et al. [17] applied the DOE method and carried out experimental measurements to create regression models describing a correlation between the tension load and the height at a given weight for two types of conveyance systems—with and without the support system. The effects of carrying idlers on energy consumption of a belt conveyor were discussed in the paper [18]. Loading of carrying idlers used in belt conveyors was investigated by the authors of paper [19], and an analysis of failures of idlers based on the measurements conducted during their operation was discussed in [20]. Sensors have become crucial components in identification of conveyor belt damage [3,12,21]. The authors in [21] developed a sensor system for the detection of stones and an automatic system for bulk materials for monitoring conveyor belt damage. Kovanic et al. proposed a method of measuring an impactor trajectory during the dynamic impact load using a laser distance sensor L-GAGE and experimental photography. This method has been experimentally tested and the results of the measurements were presented in publication [12]. Blažej et al. developed a magnetic sensor for monitoring steel-cord conveyor belts [3].

Dynamic characteristics have been investigated by many authors. For example, papers [22,23] investigated into long belt conveyance systems in terms of reducing production costs and optimise conveyor performance. Lodewijks [24] conducted the modelling of stresses in a conveyor belt during its launch and stoppage. Dynamic behaviour of a conveyance system was mathematically described by applying an analytical approach and the Finite Element Method (FEM). Authors Junxia Li and Xiaoxu Pang simulated various factors that affect longitudinal vibrations of belt conveyors and they proposed strategies for eliminating them [25]. The simulation approach proposed by those authors facilitates better accuracy of designing the dynamics of belt conveyors. The authors of papers [26,27] dealt with the stability of the motions of conveyor belts and bulk materials on long conveyor belts and proposed novel designs of actuators with the aim of reducing stresses in conveyor belts. Many of these topics address conveyance systems in terms of their dynamics, but they are not directly associated with modelling the dynamics that occur during the impact process. Modelling the dynamics of conveyance systems equipped with a support system is currently being investigated by only a few authors in their articles [28,29].

The present article, in particular, addresses the absence of research into a dynamic impact process and proposes several dynamic models while applying selected statistical methods. Within the creation of dynamic models, the research objectives were to identify and compare the amounts of absorbed energy before the occurrence of damage to the conveyor belt (puncture) which leads to belt decommissioning. The authors wanted to verify the following hypothesis: "Do the drop hammer weight, impact height, and support system absence/presence have a significant effect on a relative amount of absorbed energy before a puncture occurs?" The main goal of the research was to create a model of a correlation between the relative amount of absorbed energy and the selected parameters.

The objectives of this experimental investigation were as follows:

Identification and comparison of the amounts of absorbed energy before the occurrence of damage to conveyor belts (puncture) which leads to belt decommissioning.

Identification of the effects of parameters (drop hammer weight, impact height, support system absence/presence) on a relative amount of absorbed energy before a puncture occurs.

Creation of a model of a correlation between the relative amount of absorbed energy and the selected parameters.

2. Materials and Methods

Test Equipment

The test equipment which was used in the experiments is shown in Figure 1. A more detailed description of the equipment is presented in papers [30]. The test equipment facilitates recording the drop hammer height and magnitudes of the tension force and the impact force over time. For the purpose of our experiment, we recorded the drop hammer height (at the impact and when bouncing) every millisecond. The measurements were carried out for two different assemblies—with and without the support system. The upper limits for the drop hammer weight and impact height were determined in order to avoid destruction of the test specimen. These limits were applied in the analyses of the measurements carried out with and without engaging the support system.

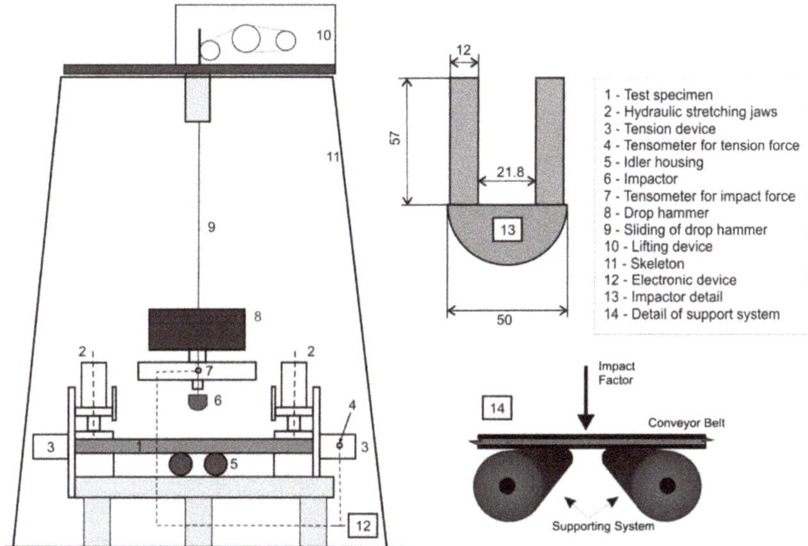

Figure 1. Test equipment scheme.

The support system consisted of two steel rollers with a diameter of 80 mm and an axial distance of 160 mm. The point of impact of the drop hammer was located between the rollers. The spherical impactor (Figure 1) simulated an impact of a bulk, brittle material. In the experiments, the simulated weights of the falling material were 50, 60, 70, 80, 90, and 100 kg. Drop heights ranged from 1 to 2.6 m, with 0.2 m increments. The investigated belt was a rubber-textile conveyor belt, type P2500 (P means the polyamide textile layer and 2500 expresses the rigidity of the conveyor belt [Nmm^{-1}]). Each specimen was sized 1400×160 mm. The specimens were prepared following the methodology described in paper [31]. The tension force applied in the experiments was 40 kN, representing 1/10 of the belt strength per millimetre of its width (manufacturer's recommendation).

3. Theory/Calculation

3.1. Absorbed Energy

With known heights to which the material bounced-back, while considering the differences in potential energies of an object prior to the impact and after the impact, it was possible to identify the amount of energy absorbed by the conveyor belt while neglecting the effects of the environment. The underlying principle is shown in Figure 2.

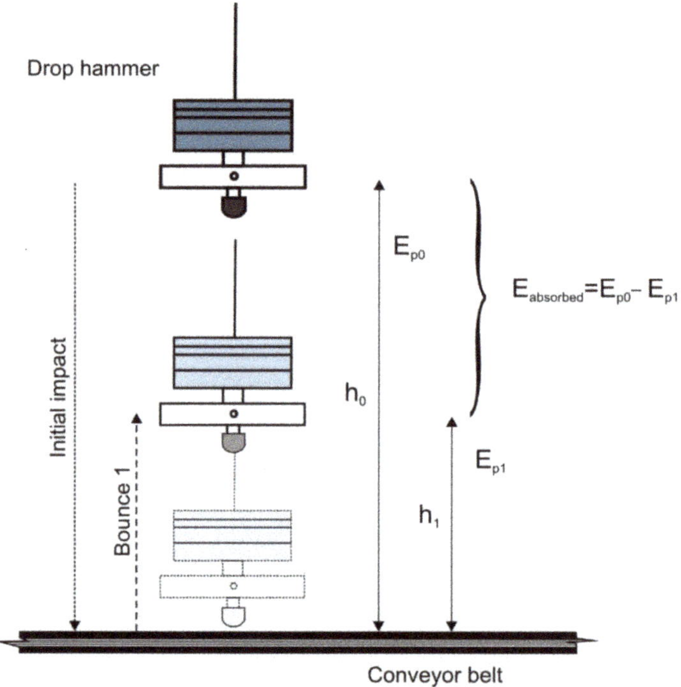

Figure 2. Impact and Bounce 1 of the drop hammer.

The absolute value of the amount of energy $E_{absorbed}$ absorbed after Bounce 1 was calculated using the following equation:

$$E_{absorbed} = E_{p0} - E_{p1},\tag{1}$$

wherein E_{p0} is the amount of potential energy prior to the first impact, and E_{p1} is the amount of potential energy after the first bounce (Bounce 1).

The amount of impact energy (in %) absorbed by the conveyor belt after the first impact was calculated using a relative amount of absorbed energy E_{relat}. The following equation was used:

$$E_{relat} = \frac{E_{p0} - E_{p1}}{E_{p0}} \times 100\% = \left(1 - \frac{E_{p1}}{E_{p0}}\right) \times 100\%.\tag{2}$$

The following equation applies to the ith bounce ($i = 1,2,\ldots$) of the drop hammer on the conveyor belt:

$$E_{relat,i} = \frac{E_{pi-1} - E_{pi}}{E_{pi-1}} \times 100\% = \left(1 - \frac{E_{pi}}{E_{pi-1}}\right) \times 100\%.\tag{3}$$

3.2. Evaluation Methods

The values were compared using the methods of statistical induction–hypothesis testing. Normality of data sets was verified using the Shapiro–Wilk test of normality. A comparison of two dependent data sets was carried out using the paired *t*-test. The null hypothesis acceptance or rejection was based on a *p*-value. In principle, if the *p*-value is lower than the significance level α, then the null hypothesis is rejected in favour of the alternative hypothesis. If the *p*-value equals to or is higher than the predetermined significance level α, then the null hypothesis is not rejected.

The correlations between the output variable and selected input variables were investigated using regression and correlation analyses. The following standard linear regression model was considered:

$$Y = \beta_0 + \sum_{j=1}^{k} \beta_j X_j + \varepsilon \tag{4}$$

wherein β_0 and β_j for j = 1,2, $\cdots$, k are the model parameters; Y is the input (dependent) variable; variables X_j, j = 1,2, $\cdots$, k represent k independent input variables; and ε is the random error. Model parameters were identified using the method of least squares.

Statistical significance of the regression model, or its parameters, was verified using the tests of statistical significance. The strengths of the effect of Y variable and the effects of k variables were expressed by means of the coefficient of multiple determination r^2. The coefficient values ranged within the <0;1> interval. In principle, as its value approaches 1, the correlation becomes stronger.

4. Results

Experimental tests were carried out with a P2500 rubber-textile conveyor belt. The measurements were carried out with two different assemblies—with and without the support system. During the experiments without the support system, no significant damage to the conveyor belt specimen was observed. With the support system engaged, destruction (punctures) of the conveyor belt specimen was observed at certain drop hammer weights and impact heights. A puncture is the type of damage that occurs to the top cover layer, the carcass and the bottom cover layer of the belt.

The input parameters were the weight *m* of the falling material and the impact height *h*.

Figure 3 shows the occurrence of punctures during the experiments. In order to facilitate a comparison of the amounts of absorbed energy with and without using the support system, a sub-set of the measurements, out of the whole set of the obtained results, was analysed for the weights ranging from 50 to 80 kg and for the impact heights ranging from 1 to 2 m at 4 consecutive bounces of the drop hammer.

The time course of the measured heights at the impact of the drop hammer onto the conveyor belt (at a certain weight *m* and a certain impact height *h*) is shown in Figures 4 and 5. The graphs illustrate the impacts of the drop hammer onto the conveyor belt with and without the support system.

The individual drop hammer bounces exhibited apparent differences in the experiments with (SS) and without using the support system (WSS).

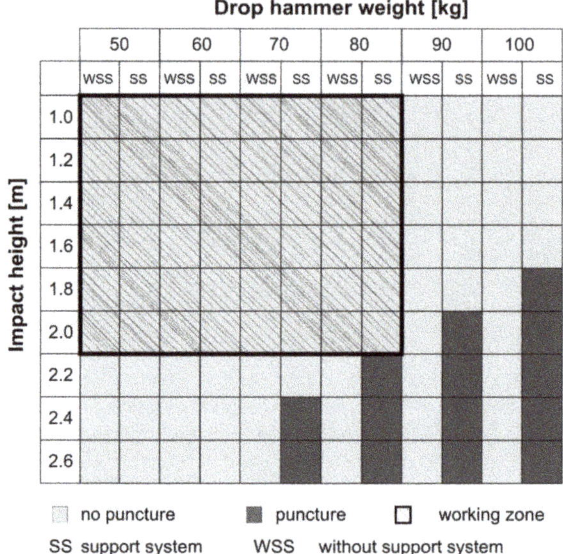

Figure 3. Puncture occurrence.

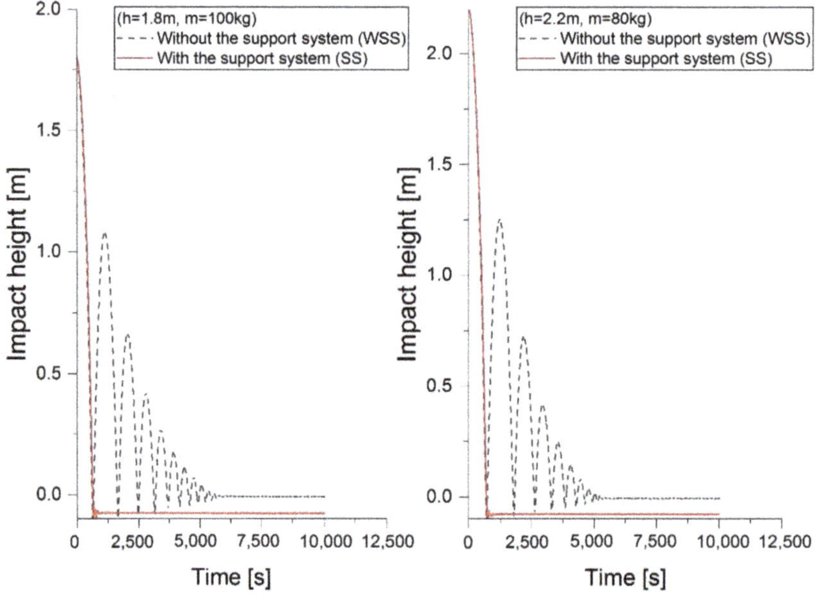

Figure 4. Time course of the drop hammer impacts with a final puncture.

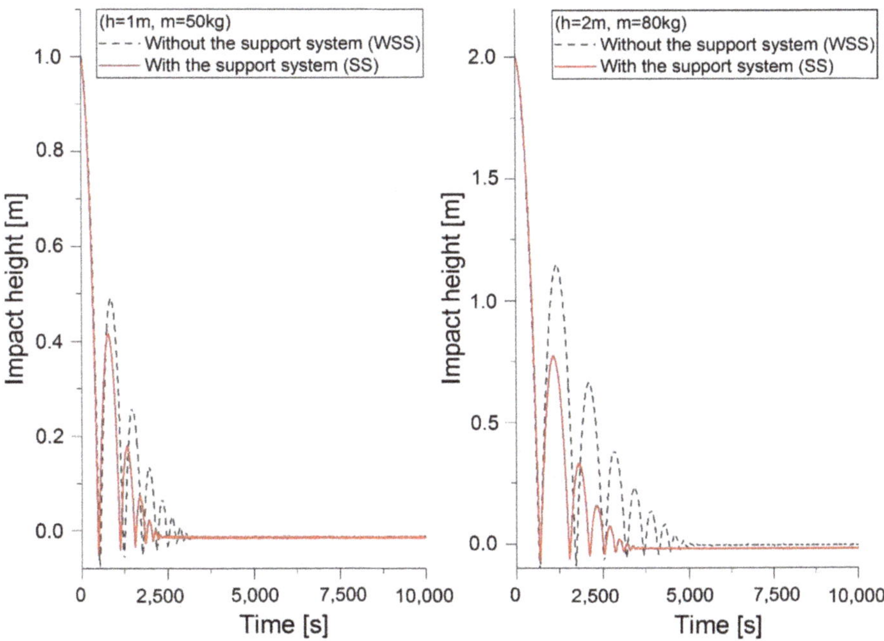

Figure 5. Time course of the drop hammer impacts.

4.1. Identification and Comparison of a Relative Amount of Absorbed Energy

Relative amounts of absorbed energy at first four bounces were expressed using Equation (3).

At drop hammer weights ranging from 50 kg to 80 kg, with 10 kg increments, impacts from six different heights were simulated. Figure 6 shows relative amounts of absorbed energy at first four bounces of the drop hammer on the conveyor belt. Two impact heights of the drop hammer, 100 cm and 200 cm, were applied to all of the tested drop hammer weights and to both system assemblies—with and without the support system.

For each drop hammer weight and each impact height, an average value of the relative amount of absorbed energy was determined. There were minimum differences between the obtained values for both impact heights, all impact weights and both system assemblies (with the support system (SS) and without the support system (WSS)) (Figure 6).

The initial comparison of the obtained values indicated a slight decrease in the values of the relative amount of absorbed energy with a growing number of bounces in both system assemblies (with and without the support system).

Average values of the relative amount of absorbed energy at the given weights, depending on whether the support system was or was not engaged, are listed in Table 1.

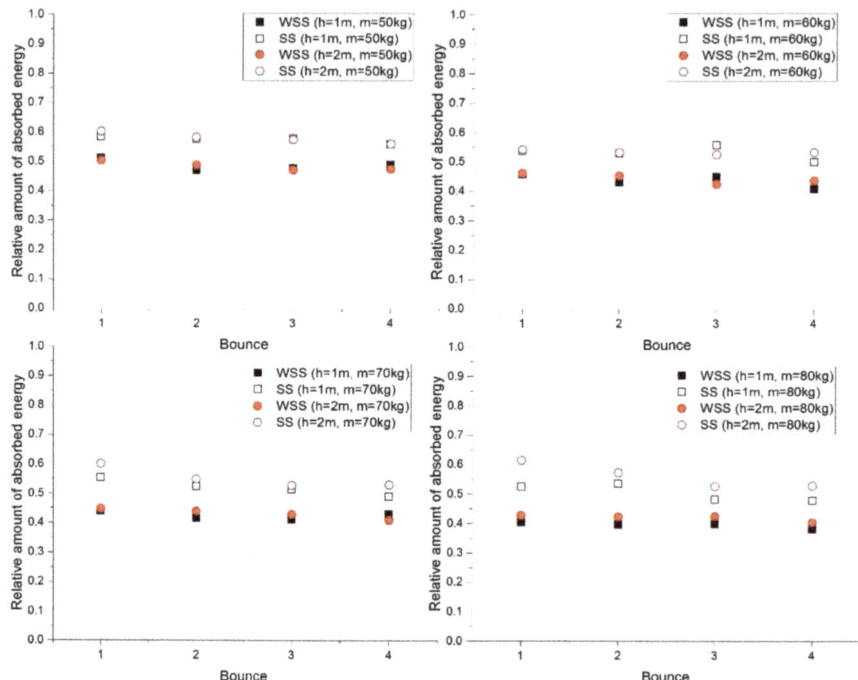

Figure 6. Relative amounts of absorbed energy.

Table 1. Average values of the relative amount of absorbed energy.

| Parameters | Drop Hammer Weight/Support System | | | | | | | |
Weight	50 kg		60 kg		70 kg		80 kg	
Support System	WSS	SS	WSS	SS	WSS	SS	WS	SS
Bounce								
Bounce 1	0.51	0.58	0.46	0.54	0.45	0.57	0.41	0.56
Bounce 2	0.48	0.57	0.44	0.53	0.43	0.54	0.41	0.54
Bounce 3	0.48	0.56	0.44	0.53	0.42	0.52	0.40	0.50
Bounce 4	0.48	0.62	0.44	0.55	0.42	0.54	0.41	0.54

The experimental measurements indicated that the effect of the impact height on the relative amount of impact energy absorbed at a given drop hammer weight was insignificant. This applied equally to both system assemblies—with (SS) and without the support system (WSS).

Nevertheless, there were some indications that a drop hammer weight affects a relative amount of absorbed impact energy. Apparently, the range of the values of the relative amount of absorbed energy is wider.

The effect of the presence of the support system on the relative amount of absorbed impact energy was evident (Figure 7, Table 1). With the support system engaged, the conveyor belt was able to absorb, for example at Bounce 1, 54% to 58% of impact energy, depending on the impact height and drop hammer weight. Without the support system, the conveyor belt was able to absorb 41% to 51% of energy at Bounce 1, depending on the impact height and drop hammer weight.

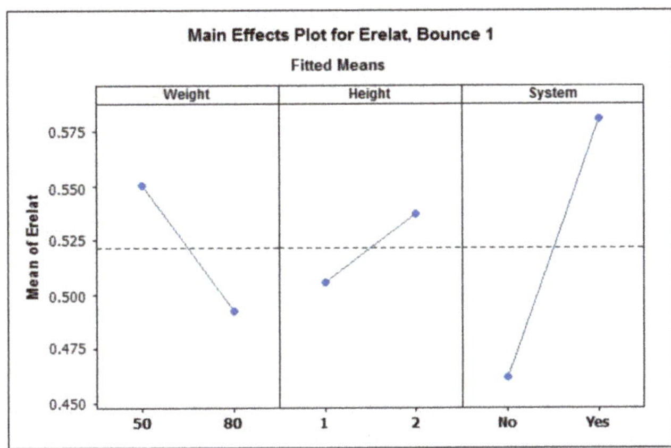

Figure 7. Graphical representation of the effects (Bounce 1).

The values obtained by measurements were compared using the testing methods. The measured values of relative absorption at the individual weights and impact heights met the normality requirement, which was verified by the Shapiro–Wilk test of normality (p-value $< \alpha$). A paired comparison of the values identified for the two system assemblies (with and without the support system) at each drop hammer weight and impact height was verified using a paired t-test. The resulting p-values obtained by the paired testing are listed in Table 2.

Table 2. Test results—paired t-test ($\alpha = 0.05$).

Impact Height [m]	Drop Hammer Weight (p-Value)			
	50 kg	60 kg	70 kg	80 kg
1.0	0.0429	0.0058	0.0006	0.0010
1.2	0.0051	0.0065	0.0004	0.0010
1.4	0.0221	0.0018	0.0006	0.0015
1.6	0.0313	0.0081	0.0014	0.0106
1.8	0.0115	0.0107	0.0005	0.0082
2.0	0.0061	0.0006	0.0004	0.0109

The results of the testing indicated that there were statistically significant differences in relative amounts of potential energy, and they depended on the conditions in which the experiments were carried out (i.e., with or without the support system, at a constant weight and impact height).

4.2. Monitoring the Effects of Selected Parameters on the Value of the Relative Amount of Absorbed Energy

The purpose of this investigation was to identify the variables (factors), or interactions between them, which have significant effects on the output variable (a value of the relative amount of absorbed energy) at the ith bounce of the drop hammer on the conveyor belt. The tests were carried out for two different system assemblies (System variable): without the support system (No) and with the support system (Yes). The minimum impact height of the drop hammer in the working zone (Height variable) was determined as 1 m, and the maximum height was 2 m. The minimum weight of the drop hammer was 50 kg while the maximum weight was 80 kg (Weight variable). The input parameters and their levels are presented in Table 3.

Table 3. Input parameters for the DOE method.

Level	Weight [kg] A	Height [m] B	System C
Lower (−)	50	1.0	No
Upper (+)	80	2.0	Yes

The effects of the main input variables (A, B, C) for all of the monitored bounces are listed in Table 4. Significance of the effects of individual variables was tested by a *t*-test and by identifying a *p*-value.

Table 4. Effects of main variables.

Bounce		Weight (A)	Height (B)	System (C)	R-Squared
1.	effect	−0.064	0.031	0.119	91.35%
	p-value	0.040	0.221 *	0.005	
2.	effect	−0.048	0.022	0.125	94.15%
	p-value	0.018	0.148 *	0.001	
3.	effect	−0.069	0.015	0.098	94.58%
	p-value	0.003	0.243 *	0.001	
4.	effect	−0.082	−0.035	0.156	95.73%
	p-value	0.005	0.070 *	0.0001	

*: *p*-value $> \alpha$, $\alpha = 0.05$.

A graphical representation of the main effects of all factors at Bounce 1 is shown in Figure 7. A positive effect value means that, as the variable shifts from the lower level to the upper level, its effect increases and so does the output value. A negative value of the effect of the drop hammer weight means that, as the weight increases, with the other conditions being constant, the output variable slightly decreases.

Results of the DOE method confirmed the conclusions made on the basis of the experimental measurements. An analysis of the results indicated that System variable (C) and Weight variable (A) had significant effects on the output variable. On the other hand, the effect of the Impact Height (C) was negligible in terms of the amount of absorbed impact energy at the given drop hammer weight, regardless of the bounce sequence number (p-value $> \alpha$).

In this case, only the Weight (A) and System (C) main variables exhibited statistically significant effects. Therefore, the model of a complete three-factor experiment was adjusted as follows:

$$E_{relat,i} = b_0 + b_1 Weight + b_2 System, \tag{5}$$

wherein $E_{relat,i}$ is the value of the relative amount of absorbed energy at i^{th} bounce and b_0, b_1, b_2 are the point estimates of the model parameters. Models (5), (6), (7), and (8) were created based on Equation (4).

The values of the parameters equalled half of the respective effect. The total informative value of the model was identified based on the coefficient of multiple determination R-squared (Table 4). High values of the R-squared coefficient at all the bounces mean that the created models explain the variability of the variables very well.

4.3. Creation of a Model of Correlations between the Relative Amount of Absorbed Energy and Selected Parameters

The purpose of this investigation was to find a regression model that would express the correlations between the relative amount of absorbed potential energy E_{relat} and four independent variables (Height, Weight, Bounce and System); the created model is as follows:

$$E_{relat} = \beta_0 + \beta_1 Height + \beta_2 Weight + \beta_3 Bounce + \beta_4 System + \varepsilon, \tag{6}$$

wherein β_0 and β_j for $j = 1, 2, 3, 4$ are the model parameters and ε is the random error.

Height was the input variable ranging from 1 m to 2 m, with 0.2 m increments and Weight ranged from 50 to 80 kg, with 10 kg increments. Bounce variable acquired four different values (1 for Bounce 1; 2 for Bounce 2; 3 for Bounce 3; and 4 for Bounce 4). System was a dichotomic variable, and it was converted into a numerical variable with two values: 1 for the support system being present (SS) and 0 for the support system being absent (WSS). The output variable E_{relat} was a continuous variable that ranged from 0 to 1.

The point estimate of the model (Model I) was as follows:

$$E_{relat} = b_0 + b_1 Height + b_2 Weight + b_3 Bounce + b_4 System. \tag{7}$$

Point estimates of the parameters, as well as their statistical significance, and interval estimates of the parameters are listed in Table 5.

Table 5. Estimated values of the regression Model I parameters ($\alpha = 0.05$).

Parameter	Estimate	Standard Error	t-Stat	*p*-Value	95%-Confidence Interval	
					Lower	Upper
Intercept	0.565	0.0117	48.311	<0.0001	0.5428	0.5890
Height	0.008	0.0045	1.674	0.096	−0.0013	0.0163
Weight	−0.002	0.0001	−13.800	<0.0001	−0.0021	−0.0016
Bounce	−0.005	0.0014	−3.845	<0.0001	−0.0080	−0.0026
System	0.101	0.0030	33.641	<0.0001	0.0955	0.1074

Results of the testing indicated that Model I was a statistically significant regression model (p-value = $2.8 \times 10^{-82} < \alpha$) with the coefficient of determination representing 0.88. Apparently, Height variable was not statistically significant (p-value $> \alpha$), while the other input variables were statistically significant, i.e., they significantly affected the output variable.

A modified regression model (Model II) was as follows:

$$E_{relat} = b_0 + b_1 Weight + b_2 Bounce + b_3 System. \tag{8}$$

Point estimates of the parameters and statistical significance of the adjusted regression model are listed in Table 6.

Table 6. Estimated values of the regression Model II parameters ($\alpha = 0.05$).

Parameter	Estimate	Standard Error	t-Stat	*p*-Value	95%-Confidence Interval	
					Lower	Upper
Intercept	0.577	0.009	60.317	<0.0001	0.5584	0.5962
Weight	−0.002	0.0001	−13.795	<0.0001	−0.0021	−0.0016
Bounce	−0.005	0.0014	−3.725	0.0003	−0.0078	−0.0024
System	0.102	0.0030	33.508	<0.0001	0.0956	0.1076

The adjusted regression Model II was statistically significant (p-value = $5 \times 10^{-83} < \alpha$) and the coefficient of determination equalled 0.88. All of the input variables were statistically significant (p-value $< \alpha$); this means that they had significant effects on the relative amount of absorbed potential energy.

A graphical representation of the empirical (real) values and the theoretical (model) values of the relative amount of absorbed energy is presented in Figure 8.

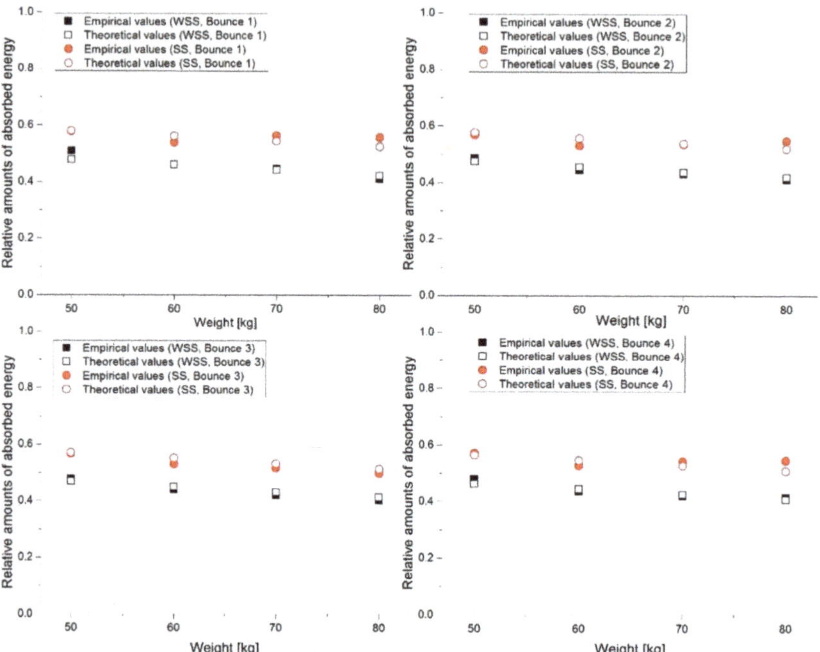

Figure 8. Empirical and theoretical values (Model II).

5. Conclusions

The purpose of this experimental research was to examine and compare the dynamic load of the P2500 rubber-textile conveyor belt. The resulting values of the relative amount of absorbed energy, obtained with and without engaging a support system, were compared. Experiments simulated various real conditions (impact of materials of various weights, falling from various heights). Conclusions that were made based on the experiments were as follows:

Within the identification and comparison of the amounts of absorbed energy before the occurrence of damage to the conveyor belt (puncture) which usually leads to belt decommissioning, it was observed that the tested conveyor belt absorbed, without suffering any damage, the energy of 1589 kJ when the support system was engaged, and 2551 kJ when the support system was absent, with a maximum impact weight of 100 kg and a maximum impact height of 2.6 m.

Based on the results of identification of the effects of the parameters (drop hammer weight, impact height and support system absence/presence) on a relative amount of absorbed energy before a puncture occurs, it was concluded that the effect of the impact height on the relative amount of absorbed impact energy at the given drop hammer weight was not strong in any of the two system assemblies, i.e., with and without the support system. However, the effect of the drop hammer weight on the relative amount of absorbed impact energy was strong. The range of the values of the relative amount of absorbed energy was apparently wider. A relative amount of absorbed energy decreased with an increasing drop hammer weight; this applied to both system assemblies, i.e., with and without the support system. Out of all the parameters, the support system presence exhibited the most evident effect on the relative amount of absorbed impact energy. The conveyor belt equipped with a support system absorbed 50% to 62% of the impact energy, depending on the drop hammer impact height and weight. Without the support system, the conveyor belt absorbed 40% to 51% of energy, depending on the drop hammer impact height and weight.

The main purpose of the research was accomplished by creating a model of correlations between the relative amount of absorbed energy and selected parameters, while the proposed dynamic model included some of the impact process parameters. The resulting regression model (8) confirmed that a drop hammer weight and presence of the support system are the parameters with a statistically significant impact on the amount of absorbed energy.

In our further research, we will carry out experimental testing of interactions between a conveyor belt and an impact bed consisting of rubber impact bars of various designs (types). We will also compare the resulting amounts of energy absorbed by the conveyor belt to those observed with a conveyor belt equipped with a conventional support system consisting of idlers. We plan to test conveyor belts of various strengths and various carcass designs.

Author Contributions: Conceptualization, M.A., D.M. and A.G.; methodology, M.A. and A.G.; software, M.A.; validation, M.A. and A.G.; formal analysis, A.G. and D.M.; investigation, M.A. and A.G.; resources, A.G.; data curation, M.A.; writing—original draft preparation, M.A., A.G. and D.M.; writing—review and editing, A.G. and D.M.; visualization, M.A.; supervision, D.M.; project administration, D.M.; funding acquisition, D.M. All authors have read and agreed to the published version of the manuscript.

Funding: This research received no external funding.

Institutional Review Board Statement: Not applicable.

Informed Consent Statement: Not applicable.

Data Availability Statement: Not applicable.

Acknowledgments: This article was prepared with the support from the project titled APVV-18-0248 "Smart belt conveyors" and the project titled VEGA 1/0168/21 "Research and application of contact and contactless methods of measuring properties of additive manufacturing products", and it was funded by National Research Agency of Slovakia, Grant No. VEGA 1/0429/18.

Conflicts of Interest: The authors declare no conflict of interest.

References

1. Saderova, J. Laboratory research of a conveyor belt with a textile carcass for mine conveying. *SGEM* **2018**, *18*, 537–544.
2. Ambrisko, L.; Marasova, D.; Grendel, P. Determination the effect of factors affecting the tensile strength of fabric conveyor belts. *Eksploat. Niezawodn.* **2016**, *18*, 110–116. [CrossRef]
3. Błażej, R.; Jurdziak, L.; Kozłowski, T.; Kirjanów, A. The use of magnetic sensors in monitoring the condition of the core in steel cord conveyor belts-Tests of the measuring probe and the design of the DiagBelt system. *Meas. J. Int. Meas. Confed.* **2018**, *123*, 48–53. [CrossRef]
4. Grujic, M.; Malindzak, D.; Marasova, D. Possibilities for reducing the negative impact of the number of conveyors in a coal transportation system. *Teh. Vjesn.* **2011**, *18*, 453–458.
5. Andrejiova, M.; Grincova, A.; Marasova, D.; Grendel, P. Multicriterial assessment of the raw material transport. *Acta Montan. Slovaca* **2015**, *20*, 26–32.
6. Grincova, A.; Andrejiova, M.; Marasova, D.; Khouri, S. Measurement and determination of the absorbed impact energy for conveyor belts of various structures under impact loading. *Meas. J. Int. Meas. Confed.* **2019**, *131*, 362–371. [CrossRef]
7. Ambrisko, L.; Marasova, D.; Cehlar, M. Investigating the tension load of rubber composites by impact dynamic testing. *Bull. Mater. Sci.* **2017**, *40*, 281–287. [CrossRef]
8. Marasova, D.; Saderova, J.; Ambrisko, L. Simulation of the Use of the Material Handling Equipment in the Operation Process. *Open Eng.* **2020**, *10*, 216–223. [CrossRef]
9. Wojtowicki, J.L.; Jaouen, L.; Panneton, R. New approach for the measurement of damping properties of materials using the Oberst beam. *Rev. Sci. Instrum.* **2004**, *75*, 2569–2574. [CrossRef]
10. Koruk, H.; Sanliturk, K.Y. On measuring dynamic properties of damping materials using beam method. In Proceedings of the 10th ASME Biennial Conference on Engineering Systems Design and Analysis, Istanbul, Turkey, 12–24 July 2010; Volume 2, pp. 127–134.

11. Komander, H.; Hardygóra, M.; Bajda, M.; Komander, G.; Lewandowicz, P. Assessment methods of conveyor belts impact resistance to the dynamic action of a concentrated load. *Eksploat. Niezawodn.* **2014**, *16*, 579–584.
12. Kovanic, L.; Ambrisko, L.; Marasova, D.; Blistan, P.; Kasanicky, T.; Cehlar, M. Long-Exposure RGB Photography with a Fixed Stand for the Measurement of a Trajectory of a Dynamic Impact Device in Real Scale. *Sensors* **2021**, *21*, 18. [CrossRef]
13. Bajda, M.; Blazej, R.; Jurdziak, L. A new tool in belts resistance to puncture research. *Min. Sci.* **2016**, *23*, 173–182.
14. Wozniak, D.; Hardygóra, M. Method for laboratory testing rubber penetration of steel cords in conveyor belts. *Min. Sci.* **2020**, *27*, 105–117. [CrossRef]
15. Grincova, A.; Berezny, S.; Marasova, D. Regresion model creation based on experimental tests of conveyor belts against belt rips resistance. *Acta Montan. Slovaca* **2009**, *14*, 113–120.
16. Amin Almasi: Don't Fall Short with Transfer Chutes. *Chem. Process* **2020**, *82*, 23–28.
17. Ambrisko, L.; Marasova, D. Evaluation the Quality of Rubber Composites using the DOE Method. *Quallity-Access Success* **2017**, *18*, 60–63.
18. Kulinowski, P.; Kasza, P.; Zarzycki, J. Influence of design parameters of idler bearing units on the energy consumption of a belt conveyor. *Sustainability* **2021**, *13*, 437. [CrossRef]
19. Król, R.; Kisielewski, W. Research of loading carrying idlers used in belt conveyor-practical applications. *Diagnostyka* **2014**, *15*, 67–74.
20. Gładysiewicz, L.; Król, R.; Kisielewski, W. Measurements of loads on belt conveyor idlers operated in real conditions. *Meas. J. Int. Meas. Confed.* **2019**, *134*, 336–344. [CrossRef]
21. Borchart, G.; Gums, B.; Kelemens, J.; Pyrcik, J. Bulk goods separation from the stream of conveyed coal. *World Min. Sufrace Undergr.* **2011**, *63*, 15–20.
22. Zhu, L.P.; Jiang, W.L. Study on typical belt conveyor in coal mine of China. *J. China Coal Soc.* **2010**, *35*, 1916–1920.
23. Wang, B.H.; Liu, J.P.; Lu, S. Analyzed development of the high-power and high-speed belt conveyor. *Min. Mach.* **2014**, *42*, 27–30.
24. Lodewijks, G. Dynamics of Belt Systems. Ph.D. Thesis, Delft University of Technology, Delft, The Netherlands, 14 October 1996.
25. Li, J.; Pang, X. Belt Conveyor Dynamic Characteristics and Influential Factors. *Shock Vib.* **2018**, *2018*, 1–13. [CrossRef]
26. Zuo, B.; Song, W.G.; Wan, L.L. Comparative of the belt conveyors starting "s curves" based on the traveling wave theory. *Min. Mach.* **2011**, *39*, 46–52.
27. Zhao, G.W. Experimental research on the rubber belts dynamic characteristics of the large belt conveyor. *Sci. Technol. Inf.* **2011**, *21*, 549–550.
28. Caban, L.; Marasova, D.; Ambrisko, L. Methodology of the Impact Process Applying the Finite Element Method. *TEM J. Technol. Educ. Manag. Inform.* **2019**, *8*, 775–781.
29. Ambrisko, L.; Frydrysek, K.; Erdeljan, D.J.; Marasova, D., Jr. Experimental research of a new generation of support systems for the transport of mineral raw materials. *Acta Montan. Slovaca* **2017**, *22*, 377–385.
30. Marasova, D.; Matiskova, D.; Andrejiova, M.; Grincova, A. Electromechanical Device with Sliding Ram for Testing Conveyor Belts (Original in Slovak). Patent Application 50037-2020, 3 September 2020. ÚPV SR.
31. Fedorko, G.; Molnar, V.; Marasova, D.; Grincova, A.; Dovica, M.; Zivcak, J.; Toth, T.; Husakova, N. Failure analysis of belt conveyor damage caused by the falling material. *Part I Exp. Meas. Regres. models Eng. Fail. Anal.* **2014**, *36*, 30–38.

Article

Inspection Robotic UGV Platform and the Procedure for an Acoustic Signal-Based Fault Detection in Belt Conveyor Idler

Hamid Shiri, Jacek Wodecki *, Bartłomiej Ziętek and Radosław Zimroz

Department of Mining, Faculty of Geoengineering, Mining and Geology, Wrocław University of Science and Technology, 50-370 Wrocław, Poland; hamid.shiri@pwr.edu.pl (H.S.); bartlomiej.zietek@pwr.edu.pl (B.Z.); radoslaw.zimroz@pwr.edu.pl (R.Z.)
* Correspondence: jacek.wodecki@pwr.edu.pl

Abstract: Belt conveyors are commonly used for the transportation of bulk materials. The most characteristic design feature is the fact that thousands of idlers are supporting the moving belt. One of the critical elements of the idler is the rolling element bearing, which requires monitoring and diagnostics to prevent potential failure. Due to the number of idlers to be monitored, the size of the conveyor, and the risk of accident when dealing with rotating elements and moving belts, monitoring of all idlers (i.e., using vibration sensors) is impractical regarding scale and connectivity. Hence, an inspection robot is proposed to capture acoustic signals instead of vibrations commonly used in condition monitoring. Then, signal processing techniques are used for signal pre-processing and analysis to check the condition of the idler. It has been found that even if the damage signature is identifiable in the captured signal, it is hard to automatically detect the fault in some cases due to sound disturbances caused by contact of the belt joint and idler coating. Classical techniques based on impulsiveness may fail in such a case, moreover, they indicate damage even if idlers are in good condition. The application of the inspection robot can "replace" the classical measurement done by maintenance staff, which can improve the safety during the inspection. In this paper, the authors show that damage detection in bearings installed in belt conveyor idlers using acoustic signals is possible, even in the presence of a significant amount of background noise. Influence of the sound disturbance due to the belt joint can be minimized by appropriate signal processing methods.

Keywords: rolling element bearing; damage; idler; belt conveyor; sound; signal processing; inspection robot

Citation: Shiri, H.; Wodecki, J.; Ziętek, B.; Zimroz, R. Inspection Robotic UGV Platform and the Procedure for an Acoustic Signal-Based Fault Detection in Belt Conveyor Idler. *Energies* **2021**, *14*, 7646. https://doi.org/10.3390/en14227646

Academic Editors: Daniela Marasová, Monika Hardygora and Mirosław Bajda

Received: 19 October 2021
Accepted: 10 November 2021
Published: 16 November 2021

Publisher's Note: MDPI stays neutral with regard to jurisdictional claims in published maps and institutional affiliations.

1. Introduction

Belt conveyors are widely recognized as interesting objects for condition monitoring [1]. There are plenty of articles focused on the diagnostics of drive units (gearboxes, pulleys) using vibration analysis or infrared thermography [2–6] or temperature [7]. The conveyor belt has been defined as one of the most expensive component in conveyor, thus various NDT techniques (image analysis, laser scanning, magnetic field measurement) have been applied [8–13].

Researches on idlers were rather focused on rolling resistance, their energy consumption, load distribution, and failure analysis until now [14–18], however, some infrared thermography applications can be found in [5,8,9], among others.

There are also other interesting research problems regarding conveyors, including destructive testing of steel-core belts, modeling of material stream behavior in a transfer point (between two conveyors), or detection of humans in harsh conditions (for the case when they are using conveyors as transport means for miners) [19–21]. However, these topics have no direct link to the predictive maintenance of such systems.

Rolling element bearings have been widely discussed in the literature [22–24]. In the case of conveyors, bearings are used in electric motors, gearboxes, pulleys, and, on a massive scale, in idlers. Mostly, vibration signals are used for fault detection [22,23,25].

Acoustic emission or just acoustic (sound) signals are rarely explored [26–31]. Delgado-Arredondo at all [32], introduced methodology-based sound and vibration signals to fault detection on an induction motor. They used Ensemble Empirical Mode Decomposition to decompose the signal. Afterward, Gabor transform was utilized to calculate the spectral content of signals in the frequency domain. Their method could detect two broken rotor bars and mechanical unbalance defects. In [33] data driven methodology to detect fault in the combustion motor were introduced based on the Wavelet Packet Transform (WPT), Principle Component Analysis (PCA), and Bayesian optimization by using the acoustic signal.

A very interesting application of acoustic-based condition monitoring using a mobile phone has been proposed in [26,27].

Bearings used in idlers are not very often discussed in the literature. The reason for that is likely the fact that vibration measurements are very difficult to perform for so many idlers. Therefore, it might be the best solution (or rather one of the most feasible solutions) to use a sound signal. Unfortunately, there are additional sources of noise and fault detection in such conditions that may be complicated to distinguish, especially if one can consider impulsive (what we will discuss later) disturbances in the acoustic signal.

An impulsive noise in signal processing procedures developed for local damage detection has been recognized as a critical challenge [34–48]. It is well known that faulty bearings produce a cyclic impulsive signal and the properties of the Signal of Interest depend on the bearing geometry, rotational speed, and fault size. There are two commonly used approaches for damage detection: searching for impulsiveness and periodicity. In the presence of impulsive disturbance, the first approach fails completely. To identify signatures in the spectrum with so-called fault frequency, one needs some preprocessing, which in practice is prefiltering used to select an informative frequency band. Again, prefiltering is mostly based on the search of impulsive behavior, which is hardly acceptable in this case. Moreover, even a classical envelope spectrum analysis or very advanced bi-frequency maps related to cyclostationary signal analysis may fail if the impulsive disturbance is really high [39–42,44,46].

As mentioned, due to the number of idlers located along the conveyor, there is a need to have a method for quick and automatic acoustic measurement and analysis. Thus, there are intensive works on inspection robots equipped with various sensors and data acquisition systems, including sound recording [5,6,8,9,49–51]. In the paper, we propose a combination of robotics inspection, acoustic data measurement, and finally signal processing for fault detection in idlers. The novel approach lies in proposing multidimensional data structure called spectral autocorrelation, which is an extension of the ordinary autocorrelation function calculated for the subsignals decomposed in the carrier frequency domain. The advantages of using such map are laid out at the final stage of result demonstration.

The acoustic signals acquired from mining machines are commonly mixed with high-energy environmental noise and interfaces from other neighboring devices. Furthermore, mining machines commonly work under time-varying speed, uncertain load conditions, and noncyclic impulsive noise that lead signals to have a complicated structure. Therefore, it is necessary to reduce this effect on the original signal by applying preprocessing methods. Adaptive mode decomposition approaches are well known and effective methods for dealing with such complex signals. They can identify the local properties of a signal with great confidence of separability. Adaptive mode decomposition approaches are well known and are effective methods for dealing with such complex signals. They can identify the local properties of a signal with great confidence of separability. Empirical mode decomposition (EMD) is a well-known adaptive mode decomposition method proposed to analyze non-stationary signals [52,53]. However, this method is suffering from mode mixing, endpoint effect, pseudo pulses, and other phenomena. Therefore, the local mean decomposition (LMD) was developed to determine the mode mixing problem in EMD. LMD utilized smoothed local means to evoke intrinsic modes from a signal instead of

Hilbert transform used in the EMD algorithm [54–56]. Consequently, the information loss prompted by the Hilbert transform can be minimized.

The paper is structured as follows: after the introduction, the experiment is described and the most problematic aspects of the data processing are indicated. After that, the key aspects of the processing methods are described in theory. Finally, the results are presented with an indication of all intermediate steps and the conclusions are formed.

2. Experiments and Data Description

Idlers are rotating elements of the belt conveyor installed to support a moving belt. It consists of a shaft, two bearings, and coating. As the number of idlers is massive in practice (see Figure 1), a quick contactless method for bearing condition evaluation is required. The experiment discussed in this paper has been done on a real industrial object during regular operation. The conveyor itself is installed in a hall encapsulating the final process of the material extraction. In this hall, the final stage of belt conveyors is operated in order to discharge the material transported from the mining pit to the appropriate compartments that are storing the material temporarily. Hence, in this hall some sections of the conveyors are transporting the material to be dropped to the silos, and some of them perform a function of the final stage of conveyor series, where the belt is finalizing the loop and returning towards the excavation area. Until now, the inspection is done in a classical way, i.e., the expert is checking the condition using their own senses, i.e., sight and hearing (see Figure 2). However, the ultimate goal of the project is to replace such "human-based inspection" with a mobile inspection robot, as shown in (see Figure 3). In this figure one can see two belt conveyors installed in parallel. At the moment of the experiment, the right conveyor was stationary, while the left one was in operation and it was a subject of the inspection. While laser scanner was configured in a way to scan both of them, the headpiece containing other measurement devices, such as cameras, was directed towards the left conveyor.

Figure 1. A general view of the belt conveyor.

Figure 2. Traditional idler inspection.

Figure 3. View of the robot during inspection.

During the experiments, a sound from each idler has been recorded, among others. The duration of the signal is c.a. 10 s, with a sampling frequency of 48 kHz. Several interesting examples have been noticed, i.e., some idlers generated a cyclic impulsive signal, which is a clear signature of faulty bearings (see exemplary signals in Figure 4). However, it has been found that for some measurements, strong impulsive disturbances appeared (see exemplary signals in Figure 5).

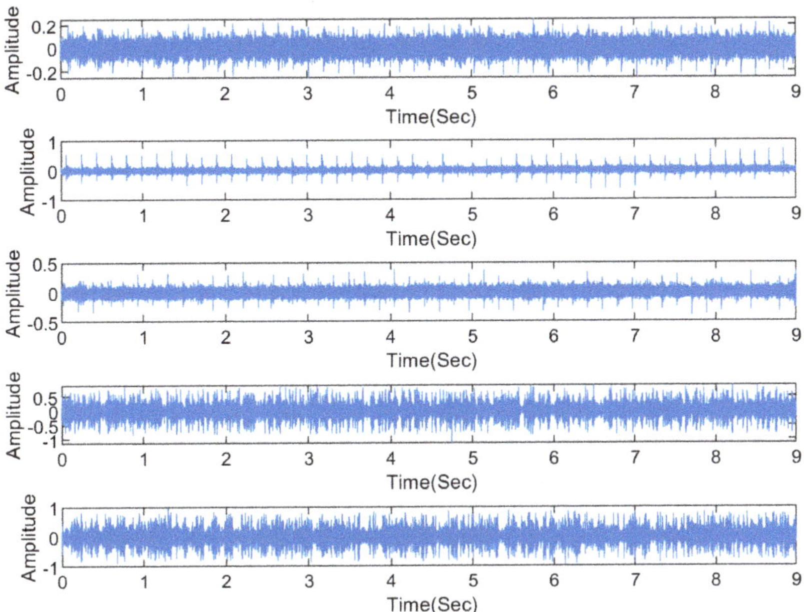

Figure 4. Selected examples of cyclic impulsive signals corresponding to faulty idlers.

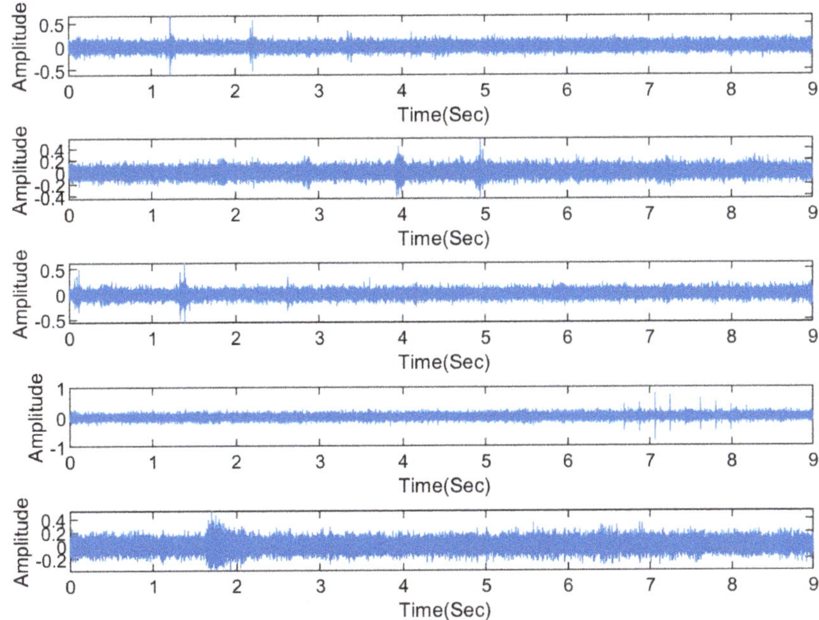

Figure 5. Selected examples of noncyclic impulsive signals corresponding to signal disturbances.

It may be concluded that several classes of signals can be distinguished in such a scenario, namely: non-impulsive signals (healthy case), cyclic impulsive signals (faulty case), noncyclic impulsive signals (or more precisely, impulsive signals that are not related to damage), and finally a mixture of cyclic and noncyclic signals. It is obvious that signals with different properties require more general, i.e., more advanced analysis. The two last cases are difficult to deal with, indeed. Such high energy impulse are represented in the frequency domain as wide-band excitation (see time-frequency representation of the exemplary signal in the Figure 6) and complicate the procedure of so-called informative frequency band detection. A similar effect can be observed in the time domain on the fourth subplot of Figure 5. The source of these noncyclic impulses (not related to damage) is a metal clip connecting two pieces of belt, which moves over the idlers and hits them with a force much stronger than a typical interaction between the belt and idler, see Figure 7. In this context it is important to know that the belt is never a single consistent loop. It is manufactured as a linear strip, which is then installed as a loop by making a connection of the ends. There are many ways to make such connection, which depends on the type of belt used in a given conveyor (steel-core belt, textile belt, etc.) and on the preferred method of making a joint in a given application (sawtooth gluing, layered gluing, multilayer gluing, thermal vulcanization of several types, various types of mechanical connections, etc.).

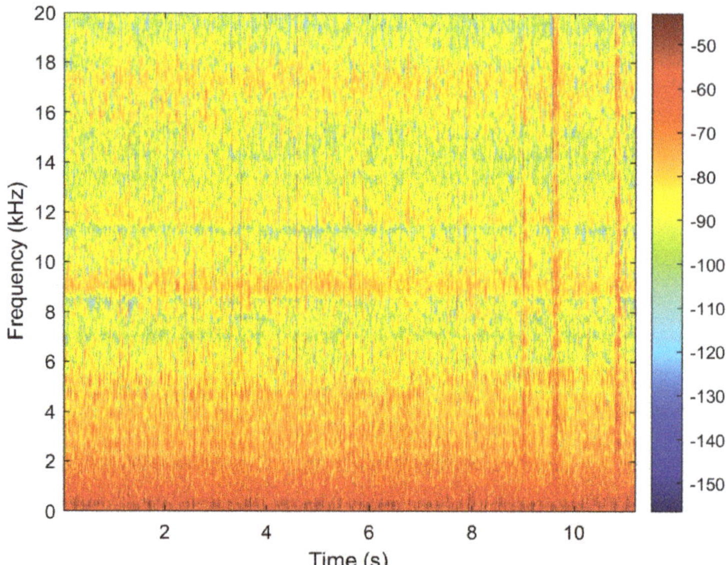

Figure 6. Time-frequency representation of the signal with impulsive disturbance.

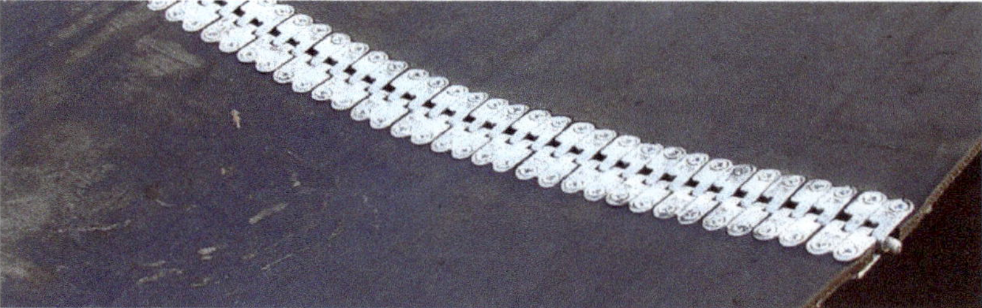

Figure 7. Belt joint using a metal clip.

3. Methodology

In this section, the key elements of the methodology are described. A general flowchart of the procedure is presented in Figure 8.

Firstly, local mode decomposition (LMD) is used as a way to denoise the input signal in a data-driven manner. Then, a spectral autocorrelation (SAC) map is calculated for the denoised signal. It allows to observe cyclic behavior in the signal with respect to the carrier frequency spectrum. In the final stage of the result section, it is shown how the usage of such an approach provides better results than using ordinary autocorrelation function (ACF). In the next step, the SAC map is spatially denoised to further enhance its quality. Finally, such an enhanced map is integrated along the frequency dimension, which provides enhanced autocorrelation function (EACF), which operates with the same concept as enhanced envelope spectrum produced when cyclostationarity maps are integrated after enhancements. In the end, EACF allows to confirm the frequency of the idler rotation, which indicates the bearing fault.

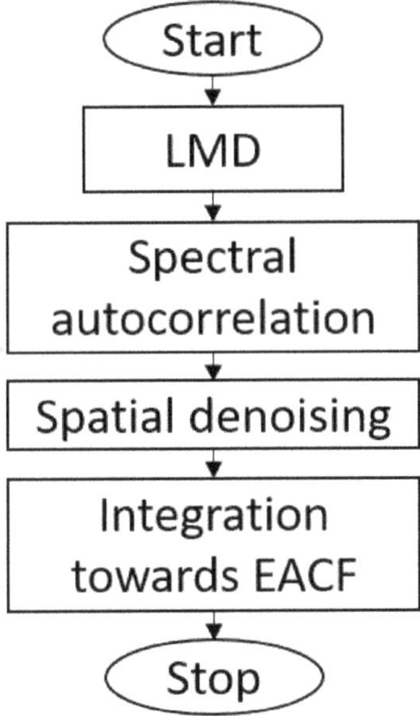

Figure 8. Flowchart of the procedure.

3.1. Preprocessing

LMD decomposes a complicated signal into a series of product functions (PFs) and a residue. Every PF is a mono-component that is produced from an envelope signal and a frequency modulated signal. In addition, A two-lever excursion algorithm is utilized to integrate the decomposition. The first step, a precise calculation of the PF, is augmented through the inner cycle. Second, in the outer cycle, the decomposition process of the signal is performed based on iterations. For implementing LMD on the signal $x(t)$, eight steps are required.

Step 1: Extract all local extrema n_i from the raw signal $x(t)$. Compute the local envelope estimate a_i and local value m_i of two consecutive extrema n_i and $n_{(i+1)}$ by utilizing (1) and (2), respectively.

$$m_i = \frac{n_{i+1} + n_i}{2} \tag{1}$$

$$a_i = \frac{|n_{i+1} - n_i|}{2} \tag{2}$$

Step 2: Make connections using direct lines between the local envelope estimate a_i and local mean values m_i.

Step 3: By applying the moving average method to smooth the local mean and envelope estimate, create the amplitude function $a_{11}(t)$ and local mean function $m_{11}(t)$.

Step 4: Calculate residue signal $h_{11}(t)$ by subtracting local mean function $m_{11}(t)$ from the raw signal.

$$h_{11}(t) = x(t) - m_{11}(t) \tag{3}$$

Afterward, calculate the frequency modulated signal $s_{11}(t)$ as follows

$$s_{11}(t) = \frac{h_{11}(t)}{a_{11}(t)}. \tag{4}$$

Step 5: To extract the envelope estimate $a_{12}(t)$ of $s_{11}(t)$, replicate steps 1–3. If the envelope function $a_{11}(t) = 1$, interrupt the process and select $s_{12}(t)$ as the first frequency modulated (FM). Otherwise, select $s_{11}(t)$ instead of the raw signal and repeat Steps 1–4 n times until the envelope function $a_{1(n+1)}(t)$ of $s_{1n}(t)$ convince $a_{1(n+1)}(t)$. The first iterative procedure can be defined as.

$$\begin{cases} h_{11}(t) = x(t) - m_{11}(t) \\ h_{12}(t) = s_{11}(t) - m_{12}(t) \\ \vdots \\ h_{11}(t) = s_{1(n-1)}(t) - m_{11}(t) \end{cases} \tag{5}$$

where

$$\begin{cases} s_{11}(t) = \frac{h_{11}(t)}{a_{11}(t)} \\ s_{12}(t) = \frac{h_{12}(t)}{a_{12}(t)} \\ \vdots \\ s_{1n}(t) = \frac{h_{1n}(t)}{a_{1n}(t)} \end{cases} \tag{6}$$

Step 6: The corresponding instantaneous amplitude of the product function can be computed as follows.

$$a_1(t) = a_{11}(t)a_{11}(t) \ldots a_{1n}(t) = \prod_{q=1}^{n} a_{1q}(t). \tag{7}$$

Step 7: Create the first product function $PF_1(t)$, utilizing.

$$PF_1(t) = a_1(t)s_{1n}(t). \tag{8}$$

In theory, $PF_1(t)$ consists of the main signal $x(t)$ oscillation information. The IA of $PF_1(t)$ is $a_1(t)$, and the IF can be computed as.

$$f_1(t) = \frac{1}{2\pi} \frac{d[\arccos(s_{1n}(t))]}{dt}. \tag{9}$$

Step 8: Calculate the residue signal $u_1(t)$. Consider $u_1(t)$ as a new signal and perform the described process k times until $u_k(t)$ does not consist of oscillation. The second iterative process can be represented as follows.

$$u_1(t) = x(t) - PF_1(t)$$
$$\vdots \tag{10}$$
$$uk(t) = u_{k-1}(t) - PF_k(t)$$

Therefore, the primary signal can be reconstructed by utilizing it as follows

$$x(t) = \sum_{p=1}^{k} PF_k(t) + u_k(t) \tag{11}$$

where k is shown the number of PFs and $u_k(t)$ is the residue signal. The LMD methods flowchart is demonstrated in the Figure 9.

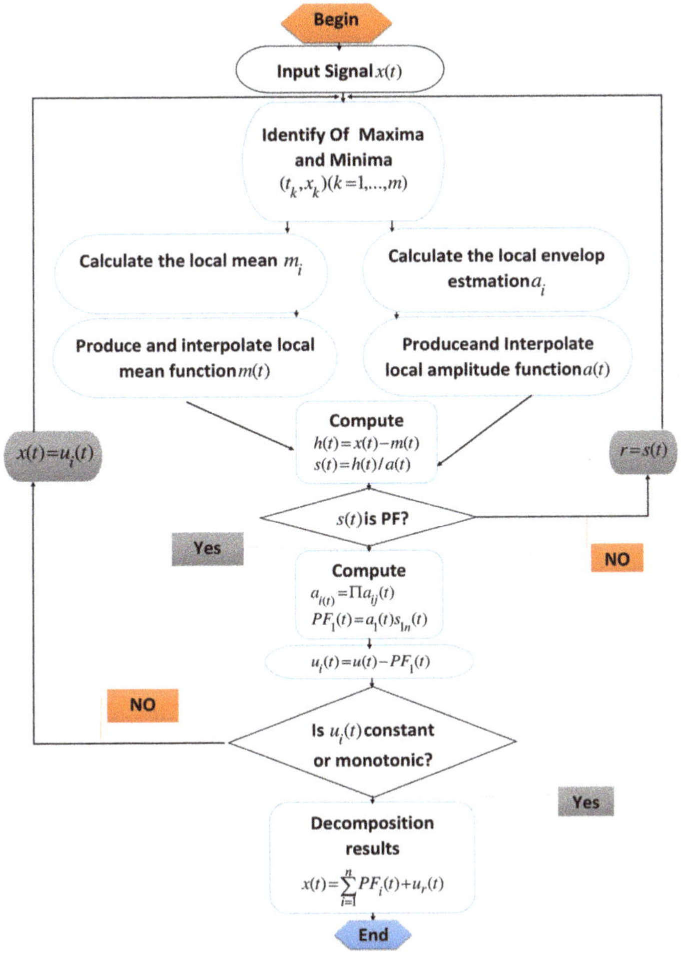

Begin

Input Signal $x(t)$

Identify Of Maxima and Minima
$(t_k, x_k)(k=1,\ldots,m)$

Calculate the local mean m_i

Calculate the local envelop estmation a_i

Produce and interpolate local mean function $m(t)$

Produce and Interpolate local amplitude function $a(t)$

$x(t)=u_i(t)$

Compute
$h(t)=x(t)-m(t)$
$s(t)=h(t)/a(t)$

$r=s(t)$

$s(t)$ **is PF?**

Yes

NO

Compute
$a_{i(t)}=\Pi a_{ij}(t)$
$PF_i(t)=a_1(t)s_{1n}(t)$

$u_i(t)=u(t)-PF_i(t)$

NO

Is $u_i(t)$ **constant or monotonic?**

Yes

Decomposition results
$x(t)=\sum_{i=1}^{n}PF_i(t)+u_r(t)$

End

Figure 9. LMD flowchart.

3.2. Spectral Autocorrelation

To detect the cyclic component present in the signal, the authors propose the use of the spectral autocorrelation method. Firstly, the signal is decomposed in terms of its carrier frequency domain. In practice, this is realized by using a filter array FA of type-1 linear-phase FIR filters with Kaiser window, and filtering the signal using the FFT-based overlap-add method [57,58].

The number of filters in the array can be set as a parameter nB and the filters have equal passband width with respect to -3 dB cutoff frequencies with the width $df = fs/(2*nB)$. -3 dB frequencies of the respective filters are also the crossover points between the neighboring filters. In practice, it means that the array consists of one lowpass filter for the frequency band $[0, df]$, one highpass filter for the frequency band $[fs/2 - df, fs/2]$, and $nB - 2$ bandpass filter for the remaining part of the Nyquist band. After filtering the input signal with this filter array, one can obtain an array of subsignals $Y^{nB \times T}$ where T is the length of a time series, which is calculated as:

$$Y_i = filter(x, FA_i), \quad for \quad i \in 1 : nB, \tag{12}$$

where the usage of operator FA_i stands for filtering the same input signal x with consecutive filters from the FA array.

In the next step, for each of the obtained subsignals, the sample autocorrelation function (ACF) is calculated. The autocorrelation itself measures the correlation between samples $y[t]$ and $y[t+k]$ of the signal y where $k = 0, ..., K$ and K is a range of the calculated autocorrelation function.

According to [59] the autocorrelation for lag k is defined as:

$$ACF[k] = \frac{c[k]}{c[0]}, \tag{13}$$

where c_0 is the sample variance of a time series and

$$c[k] = \frac{1}{T} \sum_{t=1}^{T-k} (y[t] - \bar{y})(y[t+k] - \bar{y}). \tag{14}$$

The full set of obtained ACFs is then arranged in the form of matrix $SAC^{nB \times K}$ according to the respective frequency bands of the filter that were used to obtain a given subsignal:

$$SAC_i = ACF(y_i), \quad for \quad i \in 1 : nB \tag{15}$$

3.3. Spatial Noise Modeling

To enhance the quality of the SAC map, the authors decided to introduce a preconditioning step before the actual identification step, in order for noise levels across the map to be spatially modeled and subtracted from the data for each Δf.

The spatial context is created by the analysis of each carrier frequency bin $f = [f_1, \ldots, f_{nB}]$ along the lag dimension $\alpha = [\alpha_1, \ldots, \alpha_n]$. Each vector is modeled specifically to describe the energy of the noise within this frequency band.

Considering the described modeling conditions, each vector $SAC(f_i, \alpha)$ is modeled with a dual-term exponential function using the nonlinear least squares method. The obtained parameters $a_i \in A, b_i \in B, c_i \in C, d_i \in D$ (where A, B, C, D are vectors of the parameters for the entire f domain) of the exponential function are then allowed to obtain the noise model over the entire domain α for the given f_i. Such modeled noise components are arranged into a spatial noise map N, defined as follows:

$$N(i, \alpha) = a_i \exp(b_i \alpha) + c_i \exp(d_i \alpha), \tag{16}$$

which can be subtracted from the SAC map, effectively causing its denoising:

$$SAC_d = SAC - N, \tag{17}$$

where SAC_d denotes the denoised SAC map.

4. Results

The raw input signal used in this paper is presented in Figure 10. In the first step, the signal has been denoised using the LMD method. The result of this operation is presented in Figure 11. One can notice that both the cyclic impulsive component as well as three noncyclic disturbances are more clearly visible. In general, the level of noise has been greatly reduced. To perform the analysis of cyclic behavior in the presence of large impacts originating from the mechanical joint passing over idlers, a spectral autocorrelation map has been calculated with the range parameter $K = 1$ s. Authors found out that it is not needed to set a high resolution in the frequency domain and this representation works better with much lower resolutions than one would set, i.e., for a spectrogram. Hence, the parameter $nB = 40$ turned out to be sufficient and the SAC map quality was satisfying, which can be seen in Figure 12.

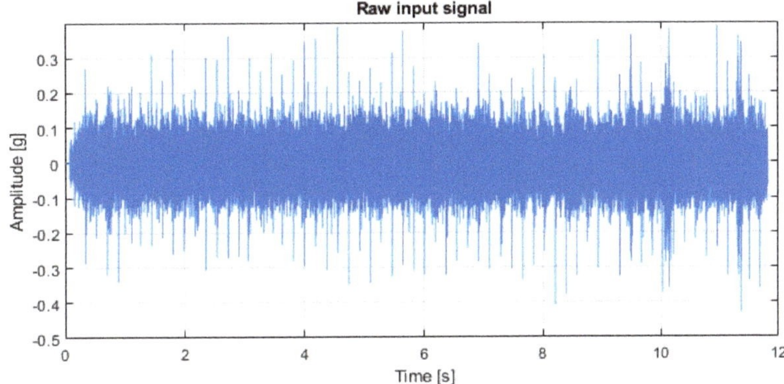

Figure 10. Raw input signal.

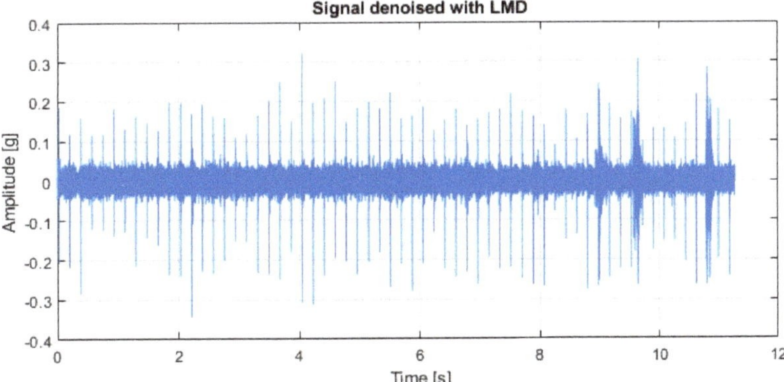

Figure 11. Signal denoised with LMD.

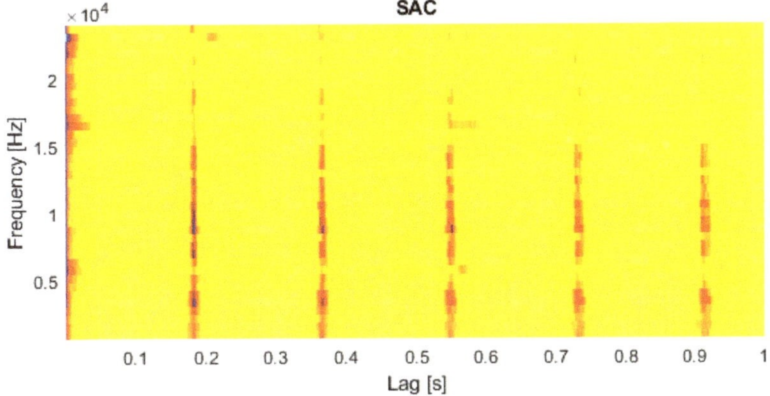

Figure 12. Spectral autocorrelation map.

In the next step, the SAC map has been cleaned up by removing the background noise profile that contains high values around the lag values of 0. To visualize this, Figure 13 presents the SAC map integrated along the frequency dimension, which displays the "equivalent autocorrelation function" with respect to the quality of the SAC map that we have at this point.

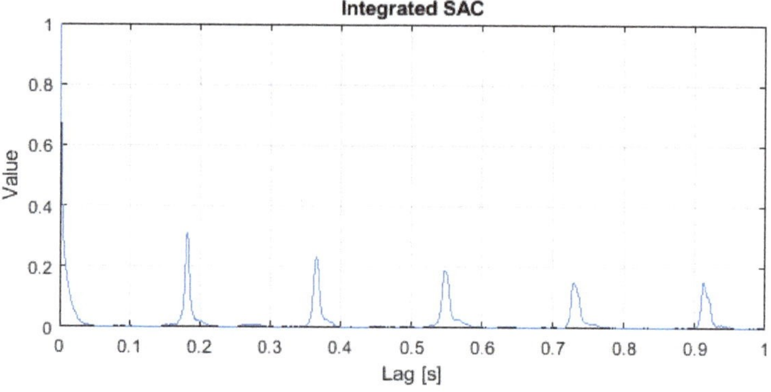

Figure 13. Integrated spectral autocorrelation map.

As expected, it is visible that the highest values are concentrated around zero lag, which interferes with thresholding or other simple detection methods. In order to remove this problem, a spatial noise model has been constructed (Figure 14) and removed from the map, which significantly improved the quality and highly simplified the subsequent steps (see Figure 15). One can notice that the vectors corresponding to the frequency bins are different from each other, which means that the unwanted background information is affecting different frequency bands with different strength. This indicates that it is reasonable to construct the model in the multidimensional domain instead of fitting a single model to the function in Figure 13. The model has been displayed in logarithmic scale to better emphasize the differences between the individual vectors.

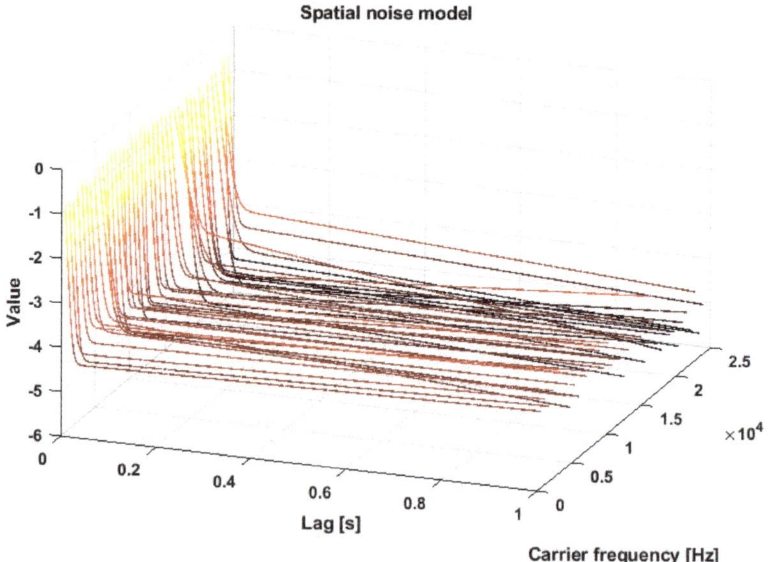

Figure 14. Spatial noise model.

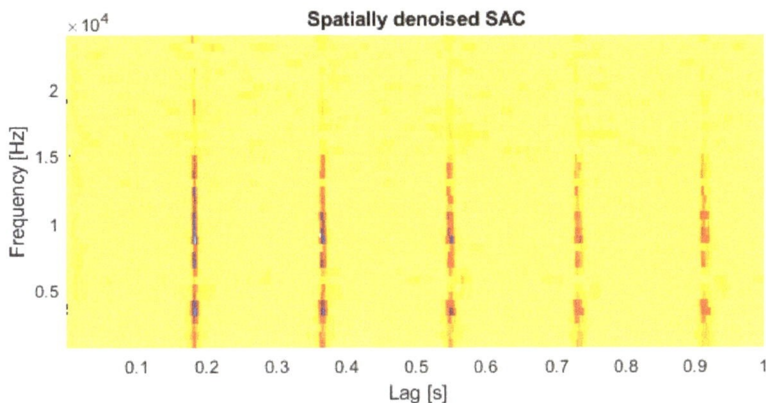

Figure 15. Spatially denoised spectral autocorrelation map.

Vectors of the denoised map have been averaged along the frequency axis, which forms a single enhanced autocorrelation function (EACF) (see Figure 16). The maximum value of this function (the value at the first peak) indicates the fundamental period of the cyclic component equal to 0.18 s, which translates to the fault frequency of 5.46 Hz.

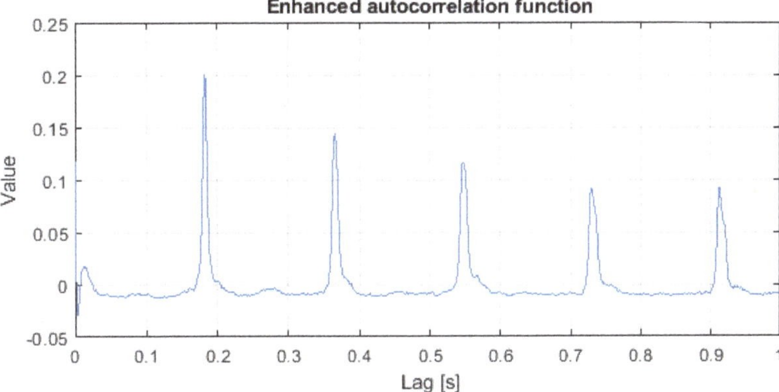

Figure 16. Enhanced autocorrelation function.

Figure 17 presents the classical autocorrelation function of the signal processed with LMD (which is the same signal that the analysis leading to EACF has been performed on). Comparison of the obtained EACF (see Figure 16) with classic ACF shows how much clarity one can achieve by choosing to perform multidimensional analysis, and how such an approach creates an opportunity to enhance the intermediate data representations leading to a better result.

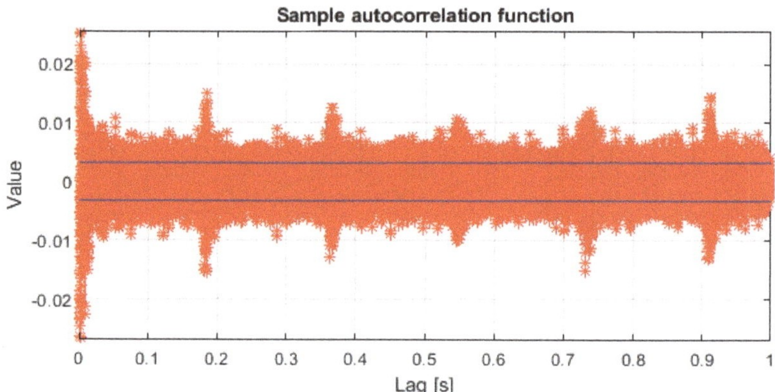

Figure 17. Classic sample autocorrelation function.

5. Conclusions

In this paper, the authors present a very promising method for damage detection in belt conveyor idlers. Based on the audio recording acquired by the mobile inspection robot, it was possible to automatically detect cyclic modulations indicating the mechanical fault of the idler. The method is especially useful because of its ability to ignore the information about random noncyclic wideband events such as mechanical connection of the belt. To achieve that, the spectral autocorrelation map was introduced as a base data structure for the analysis. Additionally, noise of the SAC map has been modeled individually with respect to the frequency domain and removed from the map. Finally, it is shown how the usage of SAC map leading to EACF provides better results than using ordinary autocorrelation function. In such scenarios, it is crucial to emphasize how simple it is to acquire data for such analysis. In our case, the mobile robot was used, but theoretically the audio signal could be recorded well enough with a simple voice recorder, or even a smartphone.

Author Contributions: Conceptualization, R.Z.; methodology, H.S. and J.W.; software, H.S., B.Z. and J.W.; validation, H.S., J.W. and R.Z.; formal analysis, J.W.; investigation, H.S. and B.Z.; resources, R.Z.; data curation, H.S. and B.Z.; writing—original draft preparation, H.S., B.Z. and J.W.; writing—review and editing, R.Z.; visualization, H.S. and J.W.; supervision, R.Z.; project administration, J.W.; funding acquisition, R.Z. All authors have read and agreed to the published version of the manuscript.

Funding: This activity has received funding from the European Institute of Innovation and Technology (EIT), a body of the European Union, under the Horizon 2020, the EU Framework Programme for Research and Innovation. This work is supported by EIT RawMaterials GmbH under Framework Partnership Agreement No. 19018 (AMICOS. Autonomous Monitoring and Control System for Mining Plants).

Institutional Review Board Statement: Not applicable.

Informed Consent Statement: Not applicable.

Data Availability Statement: Archived data sets cannot be accessed publicly according to the NDA agreement signed by the authors.

Acknowledgments: Supported by the Foundation for Polish Science (FNP).

Conflicts of Interest: The authors declare no conflict of interest.

References

1. Obuchowski, J.; Wylomańska, A.; Zimroz, R. Recent developments in vibration based diagnostics of gear and bearings used in belt conveyors. *Appl. Mech. Mater.* **2014**, *683*, 171–176. [CrossRef]
2. Błazej, R.; Sawicki, M.; Kirjanów, A.; Kozłowski, T.; Konieczna, M. Automatic analysis of thermograms as a means for estimating technical of a gear system. *Diagnostyka* **2016**, *17*, 43–48.
3. Michalik, P.; Zajac, J. Use of thermovision for monitoring temperature conveyor belt of pipe conveyor. *Appl. Mech. Mater.* **2014**, *683*, 238–242. [CrossRef]
4. Szurgacz, D.; Zhironkin, S.; Vöth, S.; Pokorný, J.; Sam Spearing, A.; Cehlár, M.; Stempniak, M.; Sobik, L. Thermal imaging study to determine the operational condition of a conveyor belt drive system structure. *Energies* **2021**, *14*, 3258. [CrossRef]
5. Carvalho, R.; Nascimento, R.; D'angelo, T.; Delabrida, S.; Bianchi, A.; Oliveira, R.; Azpúrua, H.; Garcia, L. A UAV-based framework for semi-automated thermographic inspection of belt conveyors in the mining industry. *Sensors* **2020**, *20*, 2243. [CrossRef] [PubMed]
6. Yang, W.; Zhang, X.; Ma, H. An inspection robot using infrared thermography for belt conveyor. In Proceedings of the 2016 13th International Conference on Ubiquitous Robots and Ambient Intelligence (URAI), Xi'an, China, 19–22 August 2016; pp. 400–404. [CrossRef]
7. Grzesiek, A.; Zimroz, R.; Śliwiński, P.; Gomolla, N.; Wyłomańska, A. Long term belt conveyor gearbox temperature data analysis—Statistical tests for anomaly detection. *Meas. J. Int. Meas. Confed.* **2020**, *165*, 108124. [CrossRef]
8. Szrek, J.; Wodecki, J.; Błazej, R.; Zimroz, R. An inspection robot for belt conveyor maintenance in underground mine-infrared thermography for overheated idlers detection. *Appl. Sci.* **2020**, *10*, 4984. [CrossRef]
9. Skoczylas, A.; Stefaniak, P.; Anufriiev, S.; Jachnik, B. Belt conveyors rollers diagnostics based on acoustic signal collected using autonomous legged inspection robot. *Appl. Sci.* **2021**, *11*, 2299. [CrossRef]
10. Kozłowski, T.; Wodecki, J.; Zimroz, R.; Błazej, R.; Hardygóra, M. A diagnostics of conveyor belt splices. *Appl. Sci.* **2020**, *10*, 6259. [CrossRef]
11. Trybała, P.; Blachowski, J.; Błażej, R.; Zimroz, R. Damage detection based on 3d point cloud data processing from laser scanning of conveyor belt surface. *Remote Sens.* **2021**, *13*, 55. [CrossRef]
12. Liu, X.; Pang, Y.; Lodewijks, G.; He, D. Experimental research on condition monitoring of belt conveyor idlers. *Meas. J. Int. Meas. Confed.* **2018**, *127*, 277–282. [CrossRef]
13. Liu, Y.; Miao, C.; Li, X.; Xu, G. Research on Deviation Detection of Belt Conveyor Based on Inspection Robot and Deep Learning. *Complexity* **2021**, *2021*. [CrossRef]
14. Kulinowski, P.; Kasza, P.; Zarzycki, J. Influence of design parameters of idler bearing units on the energy consumption of a belt conveyor. *Sustainability* **2021**, *13*, 437. [CrossRef]
15. Król, R.; Kisielewski, W. Research of loading carrying idlers used in belt conveyor-practical applications. *Diagnostyka* **2014**, *15*, 67–74.
16. Gładysiewicz, L.; Konieczna-Fuławka, M. Influence of idler set load distribution on belt rolling resistance. *Arch. Min. Sci.* **2019**, *64*, 251–259. [CrossRef]
17. Gładysiewicz, L.; Król, R.; Kisielewski, W. Measurements of loads on belt conveyor idlers operated in real conditions. *Meas. J. Int. Meas. Confed.* **2019**, *134*, 336–344. [CrossRef]
18. Vasić, M.; Stojanović, B.; Blagojević, M. Failure analysis of idler roller bearings in belt conveyors. *Eng. Fail. Anal.* **2020**, *117*, 104898. [CrossRef]
19. Wozniak, D.; Hardygóra, M Method for laboratory testing rubber penetration of steel cords in conveyor belts. *Min. Sci.* **2020**, *27*, 105–117. [CrossRef]
20. Doroszuk, B.; Król, R. Analysis of conveyor belt wear caused by material acceleration in transfer stations. *Min. Sci.* **2019**, *26*, 189–201. [CrossRef]
21. Uth, F.; Polnik, B.; Kurpiel, W.; Kriegsch, P.; Baltes, R.; Clausen, E. An innovative person detection system based on thermal imaging cameras dedicate for underground belt conveyors. *Min. Sci.* **2019**, *26*, 263–276. [CrossRef]
22. Rai, A.; Upadhyay, S. A review on signal processing techniques utilized in the fault diagnosis of rolling element bearings. *Tribol. Int.* **2016**, *96*, 289–306. [CrossRef]
23. Randall, R.; Antoni, J. Rolling element bearing diagnostics-A tutorial. *Mech. Syst. Signal Process.* **2011**, *25*, 485–520. [CrossRef]
24. Antoni, J.; Randall, R. The spectral kurtosis: Application to the vibratory surveillance and diagnostics of rotating machines. *Mech. Syst. Signal Process.* **2006**, *20*, 308–331. [CrossRef]
25. Roos, W.; Heyns, P. In-belt vibration monitoring of conveyor belt idler bearings by using wavelet package decomposition and artificial intelligence. *Int. J. Min. Miner. Eng.* **2021**, *12*, 48–66. [CrossRef]
26. Rzeszucinski, P.; Orman, M.; Pinto, C.; Tkaczyk, A.; Sulowicz, M. Bearing Health Diagnosed with a Mobile Phone: Acoustic Signal Measurements Can be Used to Test for Structural Faults in Motors. *IEEE Ind. Appl. Mag.* **2018**, *24*, 17–23. [CrossRef]
27. Orman, M.; Rzeszucinski, P.; Tkaczyk, A.; Krishnamoorthi, K.; Pinto, C.; Sulowicz, M. Bearing fault detection with the use of acoustic signals recorded by a hand-held mobile phone. In Proceedings of the 2015 International Conference on Condition Assessment Techniques in Electrical Systems (CATCON), Bangalore, India, 10–12 December 2015; pp. 252–256. [CrossRef]

28. Pandey, S.; Amarnath, M. Applications of vibro-acoustic measurement and analysis in conjunction with tribological parameters to assess surface fatigue wear developed in the roller-bearing system. *Proc. Inst. Mech. Eng. Part J J. Eng. Tribol.* **2021**, *235*, 2034–2055. [CrossRef]

29. Liu, X.; Pei, D.; Lodewijks, G.; Zhao, Z.; Mei, J. Acoustic signal based fault detection on belt conveyor idlers using machine learning. *Adv. Powder Technol.* **2020**, *31*, 2689–2698. [CrossRef]

30. Zhang, X.; Wan, S.; He, Y.; Wang, X.; Dou, L. Teager energy spectral kurtosis of wavelet packet transform and its application in locating the sound source of fault bearing of belt conveyor. *Meas. J. Int. Meas. Confed.* **2021**, *173*, 108367. [CrossRef]

31. Zhang, Y.; Martínez-García, M. Machine Hearing for Industrial Fault Diagnosis. In Proceedings of the 2020 IEEE 16th International Conference on Automation Science and Engineering (CASE), Hong Kong, China, 20–21 August 2020; pp. 849–854.

32. Delgado-Arredondo, P.A.; Morinigo-Sotelo, D.; Osornio-Rios, R.A.; Avina-Cervantes, J.G.; Rostro-Gonzalez, H.; de Jesus Romero-Troncoso, R. Methodology for fault detection in induction motors via sound and vibration signals. *Mech. Syst. Signal Process.* **2017**, *83*, 568–589. [CrossRef]

33. Mathew, S.K.; Zhang, Y. Acoustic-based engine fault diagnosis using WPT, PCA and Bayesian optimization. *Appl. Sci.* **2020**, *10*, 6890. [CrossRef]

34. Wang, D. Spectral L2/L1 norm: A new perspective for spectral kurtosis for characterizing non-stationary signals. *Mech. Syst. Signal Process.* **2018**, *104*, 290–293. [CrossRef]

35. Barszcz, T.; Jabłoński, A. A novel method for the optimal band selection for vibration signal demodulation and comparison with the Kurtogram. *Mech. Syst. Signal Process.* **2011**, *25*, 431–451. [CrossRef]

36. Zhao, X.; Qin, Y.; He, C.; Jia, L.; Kou, L. Rolling element bearing fault diagnosis under impulsive noise environment based on cyclic correntropy spectrum. *Entropy* **2019**, *21*, 50. [CrossRef] [PubMed]

37. Żak, G.; Wyłomańska, A.; Zimroz, R. Periodically impulsive behavior detection in noisy observation based on generalized fractional order dependency map. *Appl. Acoust.* **2019**, *144*, 31–39. [CrossRef]

38. Nowicki, J.; Hebda-Sobkowicz, J.; Zimroz, R.; Wyłomańska, A. Dependency measures for the diagnosis of local faults in application to the heavy-tailed vibration signal. *Appl. Acoust.* **2021**, *178*, 107974. [CrossRef]

39. Wodecki, J.; Michalak, A.; Zimroz, R. Local damage detection based on vibration data analysis in the presence of Gaussian and heavy-tailed impulsive noise. *Meas. J. Int. Meas. Confed.* **2021**, *169*, 108400. [CrossRef]

40. Hebda-Sobkowicz, J.; Zimroz, R.; Wyłomańska, A. Selection of the Informative Frequency Band in a Bearing Fault Diagnosis in the Presence of Non-Gaussian Noise—Comparison of Recently Developed Methods. *Appl. Sci.* **2020**, *10*, 2657. [CrossRef]

41. Hebda-Sobkowicz, J.; Zimroz, R.; Pitera, M.; Wyłomańska, A. Informative frequency band selection in the presence of non-Gaussian noise–a novel approach based on the conditional variance statistic with application to bearing fault diagnosis. *Mech. Syst. Signal Process.* **2020**, *145*, 106971. [CrossRef]

42. Wodecki, J.; Michalak, A.; Zimroz, R.; Barszcz, T.; Wyłomańska, A. Impulsive source separation using combination of Nonnegative Matrix Factorization of bi-frequency map, spatial denoising and Monte Carlo simulation. *Mech. Syst. Signal Process.* **2019**, *127*, 89–101. [CrossRef]

43. Schmidt, S.; Zimroz, R.; Chaari, F.; Heyns, P.S.; Haddar, M. A simple condition monitoring method for gearboxes operating in impulsive environments. *Sensors* **2020**, *20*, 2115. [CrossRef]

44. Kruczek, P.; Zimroz, R.; Antoni, J.; Wyłomańska, A. Generalized spectral coherence for cyclostationary signals with alpha-stable distribution. *Mech. Syst. Signal Process.* **2021**, *159*, 107737. [CrossRef]

45. Borghesani, P.; Antoni, J. CS2 analysis in presence of non-Gaussian background noise–Effect on traditional estimators and resilience of log-envelope indicators. *Mech. Syst. Signal Process.* **2017**, *90*, 378–398. [CrossRef]

46. Wodecki, J.; Michalak, A.; Wyłomańska, A.; Zimroz, R. Influence of non-Gaussian noise on the effectiveness of cyclostationary analysis – Simulations and real data analysis. *Meas. J. Int. Meas. Confed.* **2021**, *171*, 108814. [CrossRef]

47. Mauricio, A.; Qi, J.; Smith, W.A.; Sarazin, M.; Randall, R.B.; Janssens, K.; Gryllias, K. Bearing diagnostics under strong electromagnetic interference based on Integrated Spectral Coherence. *Mech. Syst. Signal Process.* **2020**, *140*, 106673. [CrossRef]

48. Yu, G.; Li, C.; Zhang, J. A new statistical modeling and detection method for rolling element bearing faults based on alpha-stable distribution. *Mech. Syst. Signal Process.* **2013**, *41*, 155–175. [CrossRef]

49. Cao, X.; Zhang, X.; Zhou, Z.; Fei, J.; Zhang, G.; Jiang, W. Research on the Monitoring System of Belt Conveyor Based on Suspension Inspection Robot. In Proceedings of the 2018 IEEE International Conference on Real-time Computing and Robotics (RCAR), Kandima, Maldives, 1–5 August 2018; pp. 657–661. [CrossRef]

50. Staab, H.; Botelho, E.; Lasko, D.; Shah, H.; Eakins, W.; Richter, U. A Robotic Vehicle System for Conveyor Inspection in Mining. In Proceedings of the 2019 IEEE International Conference on Mechatronics (ICM), Ilmenau, Germany, 18–20 March 2019; pp. 352–357. [CrossRef]

51. Garcia, G.; Rocha, F.; Torre, M.; Serrantola, W.; Lizarralde, F.; Franca, A.; Pessin, G.; Freitas, G. ROSI: A Novel Robotic Method for Belt Conveyor Structures Inspection; In Proceedings of the 2019 19th International Conference on Advanced Robotics (ICAR), Belo Horizonte, Brazil, 2–6 December 2019; pp. 326–331. [CrossRef]

52. Boudraa, A.O.; Cexus, J.C. EMD-based signal filtering. *IEEE Trans. Instrum. Meas.* **2007**, *56*, 2196–2202. [CrossRef]

53. Zheng, J.; Pan, H. Mean-optimized mode decomposition: An improved EMD approach for non-stationary signal processing. *ISA Trans.* **2020**, *106*, 392–401. [CrossRef]

54. Yongbo, L.; Shubin, S.; Zhiliang, L.; Xihui, L. Review of local mean decomposition and its application in fault diagnosis of rotating machinery. *J. Syst. Eng. Electron.* **2019**, *30*, 799–814.

55. Smith, J.S. The local mean decomposition and its application to EEG perception data. *J. R. Soc. Interface* **2005**, *2*, 443–454. [CrossRef]

56. Li, X.; Ma, J.; Wang, X.; Wu, J.; Li, Z. An improved local mean decomposition method based on improved composite interpolation envelope and its application in bearing fault feature extraction. *ISA Trans.* **2020**, *97*, 365–383. [CrossRef] [PubMed]

57. Oppenheim, A.V.; Schafer, R.W.; Buck, J.R. *Discrete-Time Signal Processing*; Pearson Education India: Upper Saddle River, NJ, USA, 1998; Volume 2.

58. Kaiser, J. Nonrecursive digital filter design using the I-sinh window function. In Proceedings of the 1974 IEEE International Symposium on Circuits & Systems, San Francisco, CA, USA, 22–24 April 1974; pp. 20–23.

59. Box, G.E.; Jenkins, G.M.; Reinsel, G.C.; Ljung, G.M. *Time Series Analysis: Forecasting and Control*; John Wiley & Sons: Hoboken, NJ, USA, 2015.

Article

Analysis of the Influence of the Type of Belt on the Energy Consumption of Transport Processes in a Belt Conveyor

Mirosław Bajda * and Monika Hardygóra

Faculty of Geoengineering, Mining and Geology, Wroclaw University of Science and Technology, Na Grobli 15 St., 50-421 Wrocław, Poland; monika.hardygora@pwr.edu.pl
* Correspondence: miroslaw.bajda@pwr.edu.pl; Tel.: +48-71-320-68-92

Abstract: Results of tests into the energy-efficiency of belt conveyor transportation systems indicate that the energy consumption of their drive mechanisms can be limited by lowering the main resistances in the conveyor. The main component of these resistances is represented by belt indentation rolling resistance. Limiting its value will allow a reduction in the amount of energy consumed by the drive mechanisms. This article presents a test rig which enables uncomplicated evaluations of such rolling resistances. It also presents the results of comparative tests performed for five steel-cord conveyor belts. The tests involved a standard belt, a refurbished belt and three energy-saving belts. As temperature significantly influences the values of belt indentation rolling resistance, the tests were performed in both positive and negative temperatures. The results indicate that when compared with the standard belt, the refurbished and the energy-efficient belts generate higher and lower indentation rolling resistances, respectively. In order to demonstrate practical advantages resulting from the use of energy-saving belts, this article also includes calculations of the power demand of a conveyor drive mechanism during one calendar year, as measured on a belt conveyor operated in a mine. The replacement of a standard belt with a refurbished belt generates a power demand higher by 4.8%, and with an energy-efficient belt—lower by 15.3%.

Keywords: steel-cord conveyor belt; energy-saving belt; belt indentation rolling resistance

Citation: Bajda, M.; Hardygóra, M. Analysis of the Influence of the Type of Belt on the Energy Consumption of Transport Processes in a Belt Conveyor. *Energies* **2021**, *14*, 6180. https://doi.org/10.3390/en14196180

Academic Editor: Sergey Zhironkin

Received: 1 September 2021
Accepted: 24 September 2021
Published: 28 September 2021

Publisher's Note: MDPI stays neutral with regard to jurisdictional claims in published maps and institutional affiliations.

1. Introduction

Research into energy-saving conveyor belts has been conducted by scientists, belt manufacturers, and by users of belt conveyors [1–3]. The choice of appropriate materials and designs in energy-saving belts has also been an object of interest at Wroclaw University of Science and Technology, where unique research methods and test rigs have been developed [4–6]. This research has been performed in cooperation with conveyor belt manufacturers and with the largest operators of such belts, i.e., with surface lignite mines, underground copper ore mines, as well as with hard coal mines. The focus has been placed mainly on steel-cord belts, as belts of this type are typically used on conveyors having greatest lengths and capacities, and such conveyors are operated in lignite mines. The drive mechanisms of such conveyors also allow greatest savings in electricity consumption. As a matter of fact, manufacturers of single-ply textile belts with aramid core and of multi-ply textile belts with polyamide-polyester core have recently started to express interest in the energy-efficiency of such belts as well, and this fact represents a promising further research option [7].

The need for investigations into lower energy consumption of belt conveyor drive mechanisms is demonstrated not only by successful and tested implementations abroad [8–10], but also, and most importantly, by the popularity of belt conveyors in the Polish mining industry. Recent long-haul conveyors incorporate unconventional technical solutions, which result from research works performed into belt conveyor resistances to motion [11,12]. In the near future, similar solutions are expected to be introduced in the Polish mining industry. In Polish copper ore mining plants, discussions have intensified on the need to replace series of short belt conveyors with several long steel-cord conveyors. Promising experiments and economic results are also reported by some hard coal mines, where rail transportation has been replaced with long belt conveyors [13].

An analysis of belt transportation systems in Polish copper ore mines shows the scale of the issue [14,15]. Currently, three mines operated by KGHM Polska Miedz S.A. have 185 conveyors (40 in the Lubin mine, 65 in the Rudna mine, and 80 in the Polkowice-Sieroszowice mine) with over 245 km of belts installed. The majority of these belts are textile multi-ply belts. The power of the drive systems installed on the conveyors is 68 MW. The Turow lignite mine operated by PGE Gornictwo i Energetyka Konwencjonalna S.A. uses more than 160 km of conveyor belts having widths from 1400 mm to 2250 mm, 70% of which are steel-cord belts. The Belchatow mine operated by the same company has over 240 km of steel-cord conveyor belts. The power of the drive mechanisms used in this mine and working in the loader–conveyor–stacker (LCS) system is more than 470 MW. Thus, mining companies are interested in lowering the operating costs of belt conveyor transportation, which depend on the costs of electricity consumption and on the operating costs of the conveyor elements, in particular belts, idlers, and drive mechanisms.

Considering the application range of belt conveyors in Polish lignite mines and the number of steel-cord belts installed on these conveyors, electricity savings from the drive mechanisms seem vast. The scale of such savings is currently being analyzed at the Belchatow mine [16,17].

This article presents the results of comparative research into the energy-efficiency of steel-cord conveyor belts. It describes the impact of replacing traditionally used belts with energy-efficient belts on the lowering of the power demand from belt conveyors. This impact is shown using the example of a conveyor operated in the Belchatow mine. The tests were performed with steel-cord belts as test objects, as this type of belts dominates in lignite mines. However, many other types of conveyor belts exist, which are commonly used in other branches of industry [18,19].

2. Materials and Methods

2.1. Belt Indentation Rolling Resistance

When a viscoelastic belt rolls on idler sets, a part of its energy is transformed (dissipated). The increase of the force in the belt required to overcome the energy losses on an individual idler set is defined as the belt indentation rolling resistance, or the rolling resistance of the belt on idlers. In a model-based approach, it is the resistance that accompanies the movement of a rigid cylinder (idler) rolling on a deformable surface (belt) [20].

The identification of the belt indentation rolling resistance was attempted with the use of theoretical data [21]. Furthermore, complicated experimental tests were performed on a conveyor and their results served to define coefficients, which additionally allowed for how rolling resistances are influenced by the belt trough, the belt design, and the force in the belt, as well as by the degree to which the belt is loaded with the transported material [22,23].

In publications [24,25], rheological models were used to derive formulas, which were for the first time based on physical parameters, including the viscoelastic properties of the belt, in such a way that they can be applied to any type of belt. Viscoelastic properties describe the behavior of elastomers at a defined load, providing a nonlinear relationship between the load and the deformation, as well as between the load and the deformation velocity. Experimental tests [26] allow accounting for the influence of certain conveyor operating parameters, i.e., belt velocity and ambient temperature, on belt rolling resistances, which is important, as conveyor belts work in extremely different climate conditions. The previous publications are based on both theoretical models and laboratory tests. The missing element was for these results to be confirmed in actual conveyor operating conditions [27]. Extensive research on how the properties of belts and their bottom covers influence belt indentation rolling resistances was presented in [28]. The tests included materials used in the bottom covers of conveyor belts made of various rubber types and with various additions of soot. They were performed in a laboratory, on a dedicated test rig in a climatic chamber, at uniform load, velocity, idler diameter, and cover thickness. The lowering of ambient temperature has been observed to have a negative influence on the indentation rolling resistances of the investigated belts.

An increase of belt indentation rolling resistances due to increasing belt loads is an important design factor in the case of long belt conveyors. Publication [29] describes a dedicated test rig for measuring belt indentation rolling resistances. The measurements can be performed based on belt load, belt velocity, idler diameter, ambient temperature, and type of rubber mixtures used in belt covers.

The above-discussed publications have become a basis for introducing national standards describing methods for measuring belt indentation rolling resistances [30–32]. In both cases, the rolling resistances were identified with the use of belts closed in loops. In order to reduce the test costs related to the significant length of the belt test sections, which need to be additionally spliced in loops, a small-scale rolling resistance test method has been proposed [33].

The literature also proposed variable velocity adjustment as a means for improving the operating effectiveness of belt conveyors [34]. A reduction of the electric energy consumption in transportation systems comprising belt conveyors can also be achieved by an optimal use of the operating schedule of such systems [35–41].

The phenomena occurring in the belt during the implementation of the propulsion process are described in the measurement methodology of other researchers [42], and make it possible to determine the energy consumption of the implemented processes.

Results of tests into the energy-efficiency of belt conveyor transportation systems [43–45] demonstrate that the energy consumption of their drive mechanisms can be limited by lowering the main resistances in the conveyor. Main resistances are related to the movement of the belt on the idler sets and are a sum of components depending on belt properties. They include belt indentation rolling resistance and belt flexure resistances. The remaining components of main resistances are represented by resistances independent of the belt properties. They include idler rotational resistances, material flexure resistance and sliding resistance of the belt on idlers. An analysis of belt movement on idler sets indicates that belt indentation rolling resistance and belt flexure resistance are significantly influenced by the properties of the conveyor belt, which account for approximately 60% of the main resistances in the conveyor. A vast majority of the main resistances (approximately 80%) is in turn represented by belt indentation rolling resistances. They can be minimized with adequately selected bottom covers of the conveyor belt. Belt indentation rolling resistance most significantly affects the electricity consumption level of the belt conveyor. Therefore, with appropriate bottom covers, belts can be designed to have reduced rolling resistances and thus the electric energy consumption of conveyor drive mechanisms can also be reduced. As a result, such a conveyor belt becomes an energy-saving belt.

This article presents a dedicated test rig and a practical method for conveyor belt tests with respect to identifying unit indentation rolling resistances. Such phenomena as contamination on the belt, belt wear and tear, abrasion of the covers, or belt cuts, which all occur frequently on the belt conveyors operated in a mine, are not observed during laboratory tests and therefore their influence was not accounted for in the measurement results.

2.2. Test Rig

The simplest and also the most accurate method for testing belt indentation rolling resistances consists in performing measurements according to the schematic diagram of Figure 1.

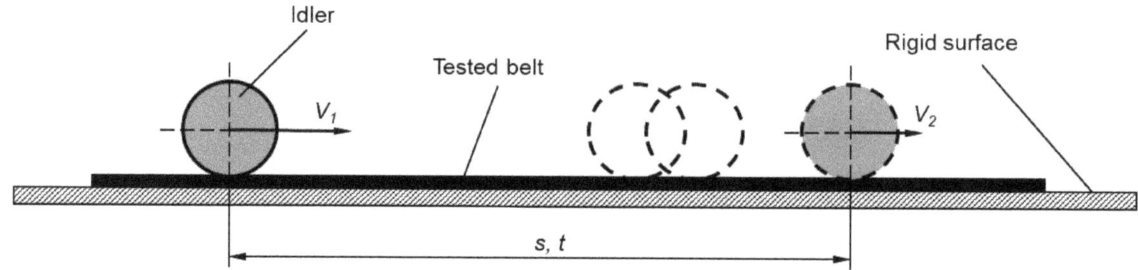

Figure 1. Schematic diagram illustrating measurements of the resistance of an idler rolling on the conveyor belt.

The principle behind this measurement is as follows. A conveyor belt having a width B is placed on a rigid and even surface, with the bottom cover facing upwards. An appropriate idler having a mass m and a moment of inertia I_r, normally engaged with such a belt on an actual conveyor, is rolled on the belt with a velocity V_1. Due to the rolling resistances, the idler rolls on the belt with a negative acceleration, losing its velocity and reaching a velocity value V_2. If the time t required by the idler to travel a distance s is measured, the result can serve to calculate the negative acceleration (the deceleration) of the idler, and subsequently, by following relationship (1), to find the unit indentation rolling resistance W_e:

$$W_e = \frac{1}{b}\left[\left(m + \frac{I_r}{r^2}\right)a\right],\tag{1}$$

Symbols used in Equation (1) represent:

W_e—unit indentation rolling resistance, N/m (the contact zone between the idler and the belt is 1 m);

m—mass of the idler, kg;

I_r—moment of inertia of the idler, kg·m^2;

r—radius of the idler, m;

a—deceleration, m/s^2.

However, the above-described method has some limitations, which prevent its practical usage. These include the following points:

- The weight of the idler is relatively small compared with the mass of the belt and the carried material. Loading a single idler with an additional force required to obtain a desired pressure is practically impossible. The idler mass cannot be simply increased, as this would entail increasing its diameter and the moment of inertia. In effect, the idler parameters would be significantly modified in comparison to typical idlers used on an actual conveyor.
- The initial velocity V_1 of the idler rolling on the belt should be approximately 3.5–6.5 m/s, so that it adequately represents the actual velocities of steel-cord belts operated on conveyors in surface lignite mines. The idler should roll on the belt along a distance of more than ten meters, so that the difference between V_1 and V_2 is significant and allows the required measurement accuracy.
- The measurement can also be performed until V_2 reaches a velocity equal to 0. In such a case, the belt specimen needed in the test should be more than 20 m in length. The belt manufacturing technology, the dimensions of vulcanization presses and the price of materials cause the cost of such a specimen to be very high and in fact prohibitive to potential clients. In the case when $V_2 = 0$, the measure in the test is the distance s traveled by the idler until it stops. An idler rolling on a belt with reduced rolling resistance will travel a longer distance in comparison to a belt having higher rolling resistance values.

Considering the above, ensuring that an idler can travel along a belt on a straight path is practically impossible. Importantly, each deviation from the rectilinear movement path generates additional resistances and consequently could result in the idler leaving the belt. Taking into account the above limitations, a decision was made that the tests would be performed with the use of two idlers positioned in a dedicated metal frame. Owing to such a solution, the idlers can travel along the belt in a straight line. The carriage frame can be additionally loaded in order to exert a desired pressure force of the idlers on the belt. A limitation of the frame-loading option is that in such case calculations of the indentation rolling resistances must include the rotational resistances of idler bearings. These resistances can be measured, however. In order to minimize the dimensions of the test specimens, a decision was made that the idlers would roll on a belt placed on a rigid and slightly inclined surface.

Design Concept

The following assumptions dictated the design and preparation of the test rig for testing the conveyor belt indentation rolling resistance:

1. The width of the tested belt specimen should be 0.5 m.
2. The length of the test specimen should be approximately 7.5 m.
3. The minimum length of the measurement section should be 5 m.
4. The tested belt should be supported along its entire length and width by a flat and rigid surface.
5. The tests should be performed with the use of two idlers secured in a rigid frame, which can be additionally loaded. It was also assumed that the outer diameter of the idler should correspond to the diameter of idlers used in the transportation of overburden in lignite mines and should be 190 mm, while the length of the idler coat should be 630 mm. The idlers should have a defined:

   - Moment of inertia Ir, kg·m^2;
   - Rotational resistance W_k, N.

6. The idler frame should be ready to accommodate weights in order to allow the idlers to exert a desired pressure against the belt.
7. An initial velocity should be given to the idler frame with the use of an inclined plane having an inclination angle of approximately 65°. The length of this plane should enable such a carriage to obtain a maximum velocity V_1 = 5 m/s with an option to adjust a lower velocity.
8. In order to reduce the distance traveled by the carriage on the belt, its measurement section should be positioned at an ascending angle of approximately 5°.
9. The test rig should be equipped with adjustable clamps allowing the proper positioning of both the raceway and the inclined plane, so as to ensure that the carriage travels in a straight line along its entire path on the tested belt.
10. The required weight of the loading carriage to be rolled on the belt installed in the test rig was calculated based on an average annual load on three-idler sets, which work with a conveyor belt of 1800 mm in width, have a trough angle of 45°, and are spaced at 1.2 m. The calculations were performed with the following parameters:

    - The nominal cross-sectional area of the transported material on the three-idler set is 0.425 m^2,
    - The bulk density of the transported overburden is 1600 kg/m^2,
    - The average annual load on the conveyor is equal to 40% of the belt nominal load, based on data provided in [46],
    - The belt width on the test rig is 0.5 m,
    - The number of idlers on the loading carriage is 2.

    The weight of the loaded carriage, as calculated based on the above data, should be 234 kg.
11. The travel time t of the carriage along a defined distance s should be measured with the use of tachometer probes, with an accuracy of up to 0.001 s. The tachometers should be secured in dedicated stands fixed to the ground in such a way that they

are isolated from the vibration of the test rig. In order to increase the measurement accuracy, it is recommended to use three such tachometer probes.

12. The belt ends should be secured in clamps that allow the belt to be tensioned.

Figure 2 schematically shows a test rig according to the assumptions described in points 1–12.

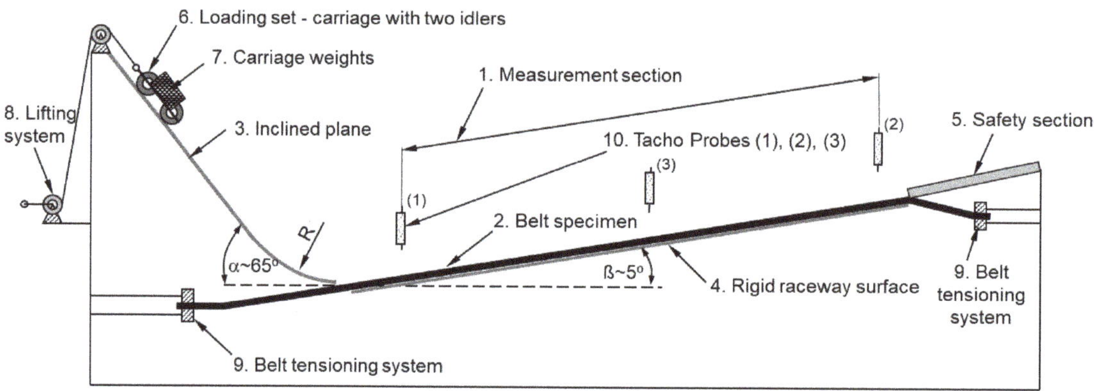

Figure 2. Schematic representation of the test rig for measuring belt indentation rolling resistances.

The symbols shown in Figure 2 are as follows:
1—measurement section;
2—belt specimen;
3—inclined plane used to accelerate the carriage with two idlers;
4—rigid raceway surface;
5—section used to decelerate and stop the idler set;
6—loading set—carriage with two idlers;
7—carriage weights;
8—system for lifting the carriage;
9—belt tensioning system;
10—tachometer probes Tacho 1, Tacho 2 and Tacho 3.

13. The belt unit indentation rolling resistance W_e on the test rig according to the schematic representation in Figure 2 should be calculated following Equation (2). A parameter W_p present in Equation (2) is related to the movement in height, as the carriage rolls on the belt arranged at an angle β with respect to the flat surface. This parameter is described by Equation (3):

$$W_e = \frac{1}{2b}\left[\left(m + 2\frac{I_r}{r^2}\right) \cdot a - W_k - W_p\right], \tag{2}$$

$$W_p = m \cdot g \cdot \sin\beta, \tag{3}$$

Symbols used in Equations (2) and (3) represent:
W_e—belt unit indentation rolling resistance, N/m;
W_k—sum of the dynamic rotational resistances of the two idlers, N;
W_p—carriage lift resistance, N;
m—mass of the carriage, kg;
a—deceleration, m/s^2;
I_r—moment of inertia of the idler, kg·m^2;
r—radius of the idler, m;
g—acceleration of gravity, m/s^2;
β—angle of the measurement section of the inclined plane;

B—belt specimen width.

Figure 3a shows the test rig for identifying belt indentation rolling resistances designed and constructed according to the recommendations listed in points 1–13. Figure 3b, on the other hand, shows the loading set comprising two idlers secured in a metal frame loaded with additional weights (the loading carriage).

Figure 3. The test rig (**a**) and the loading carriage (**b**).

2.3. Test Object

The tests covered two conveyor belts used on conveyors operated in Polish surface lignite mines and three energy-saving belts with reduced rolling resistances. They are steel-cord belts type ST 3150 (cord diameter approximately 7.4 mm) having a tensile strength of 3150 kN/m. The test specimens were cut from conveyor belts 2250 mm in width (type B-2250). The test specimens are designated as follows:

- 1-ST, standard belt commonly used in Polish lignite mines.
- 1-ST-R1, standard belt after first refurbishment (regenerated).
- 2-EO, energy-saving belt with a textile breaker in the top and bottom covers located 2 mm from the steel cords.
- 3-EO, energy-saving belt with a textile breaker in the top and bottom covers located 2 mm from the steel cords.
- 4-EO, another energy-saving belt with a breaker formed of textile cords in the top and bottom covers located 2 mm from the steel cords.

The belt specimens had the following dimensions: length—7500 mm, width—500 mm, thickness—29.5 mm. The thicknesses of the bottom covers were identical for all belts and were approximately 8 mm. Similarly, the thicknesses of the top covers and the cores were identical for all belts and were approximately 14 mm.

Table 1 lists the following physical and mechanical parameters of the rubber covers (the top and the bottom covers) for the above belts: tensile strength, elongation at break, rubber abrasiveness and hardness.

Table 1. Physical and mechanical parameters of the rubber covers.

Belt Symbol	1-ST	1-ST_R1	2-EO	3-EO	4-EO
Covers	Top/Bottom	Top/Bottom	Top/Bottom	Top/Bottom	Top/Bottom
Tensile strength, MPa	25.7/22.1	18.9/16.6	22.0/20.4	25.4/25.6	26.2/20.3
Elongation at break, mm	590/490	588/695	491/491	567/440	575/447
Abrasion resistance, mm^3	89/108	144/155	98/71	90/108	107/91
Hardness, °Sh A	62/63	64/59	64/63	62/68	61/66

The belts are types commonly used in Polish lignite mines for transporting overburden to the waste heap and coal to the power plant. Adding reinforcement in the bottom cover improves the rolling parameters of the belt on the idlers (reduced indentation of the idler is observed). On the other hand, adding a textile reinforcement in the top cover improves the resistance of the belt to failures in normal operation, such as punctures, longitudinal cuts, and tears.

2.4. Test Method

The measurement was performed as follows. The belt specimen 500 mm in width and 7500 mm in length is secured in clamps and tensioned, ensuring its stability. The belt specimen is positioned on a rigid, metal surface, with the bottom cover facing upwards. The loading carriage is lifted on an inclined plane to an appropriate height, so that it can travel with a set velocity V_1 in the first velocity measurement point in which velocity is measured with the use of the tachometer probe. After the release system is triggered, the loading set with two idlers rolls on the bottom cover of the belt, and three tachometer probes record the time t needed by the carriage to move along the belt. Prior to the measurement, the specimen was conditioned at an appropriate temperature. The measurements were performed in temperatures both above and below zero degrees Celsius. The temperature of the belt was measured at several locations along its length. The measurement was performed in the conditioning temperature. Nine measurement points were recorded. Figure 4 shows an illustrative record of the carriage travel time.

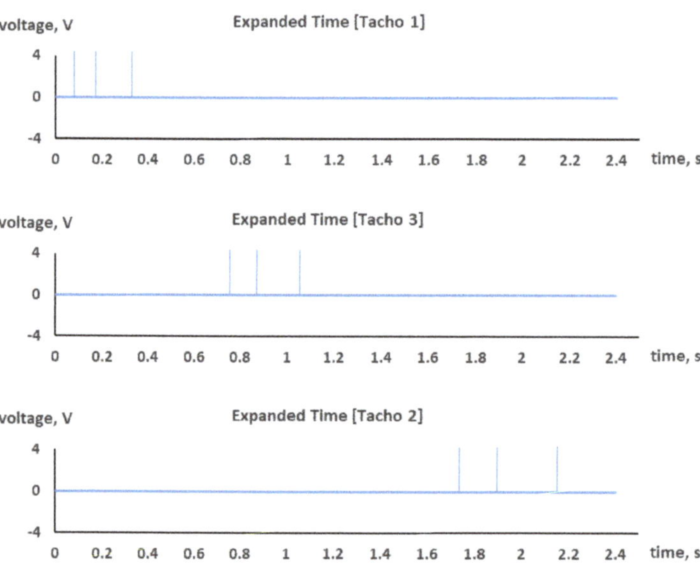

Figure 4. Time needed by the carriage to travel along the belt, as recorded by the Tacho 1, Tacho 2 and Tacho 3 probes.

The measurement results served to plot a graph of the distance traveled by the carriage in a function of time $s = f(t)$. The results were approximated with Equation (4):

$$s = \frac{a}{2}t^2 + v_0 t + s_0, \tag{4}$$

where:

a—carriage deceleration in uniformly retarded motion, m/s^2;

t—travel time of the carriage on the belt, s;

v_0—carriage initial velocity, m/s;

s_0—free term.

The value of deceleration a found from Equation (4) was used to calculate belt unit indentation rolling resistance W_e in accordance with Equation (2).

2.5. Research Scope

The laboratory tests included the identification of unit indentation rolling resistances for 5 steel-cord conveyor belts. The tests were performed with the use of two standard load-carrying idlers working on a conveyor with an 1800 mm wide belt (type B-1800). The idlers had the following parameters:

- Diameter $D = 190$ mm;
- Coat length $l = 630$ mm;
- Moment of inertia $I_r = 0.173$ kg·m^2;
- Rotational resistance of a single idler $W_k = 4.38$ N;
- The belt was loaded with unit forces 2.3 kN/m, which corresponded to the weight of the loading set. The calculations of the required carriage weight on the test rig were performed on the basis of the guidelines presented in point 10 of Section 2.2,
- Initial carriage velocity approximately 3.5 m/s;
- Measurement section length 5.20 m;
- The test temperature corresponded to the conditioning temperature.

Belt indentation rolling resistances depend on the ambient temperature, and therefore, in order to identify these relationships, the tests were performed at positive temperatures of 23.0 ± 2.0 °C and 7.5 ± 2.0 °C and at a negative temperature of -9.0 ± 2.0 °C. The temperature was measured on the bottom cover of the belt.

The scope of research into the indentation rolling resistances of standard and energy-saving belts includes a comparison of:

- the influence of bottom cover parameters on indentation rolling resistances;
- the influence of bottom cover design on indentation rolling resistances;
- the influence of bottom cover refurbishment on indentation rolling resistances.

The tests were performed for one specimen of each belt type, and the carriage traveled along the belt in ten repetitions. The obtained indentation rolling resistance values were compared with the results obtained for the standard 1-ST belt.

3. Results

Tables 2 and 3 show, respectively, values representing mean carriage travel times on belts at the positive temperatures of 23.0 ± 2.0 °C and 7.5 ± 2.0 °C. Table 4, on the other hand, shows values representing mean carriage travel times on belts at the negative temperature of -9.0 ± 2.0 °C. The tables contain mean carriage travel times t along a distance s on the belt, as recorded by the three tachometer probes. In the first research stage, the carriage travel times were measured at the beginning and at the end of the measurement section with the use of two tachometer probes (Tacho 1 and Tacho 2). However, because of the nonlinear character of the phenomena occurring in the belt as the idlers roll on it, an additional third probe (Tacho 3) was introduced in the middle of the belt length. As a result, nine measurement points were defined, which allowed Equation (4) to be fitted with more precision.

Table 2. Belt test results at a temperature of 23.0 ± 2.0 °C.

Rolling Distance of the Carriage on the Belt s (m)		Probe Tacho 1			Probe Tacho 2			Probe Tacho 3		
		0	0.273	0.772	2.175	2.448	2.947	4.389	4.662	5.161
	1-ST	0	0.0833	0.2295	0.6870	0.7849	0.9657	1.5716	1.7071	1.9738
Mean travel time	1-ST-R1	0	0.0831	0.2302	0.6909	0.7894	0.9716	1.5859	1.7255	1.9966
of the carriage on	2-EO	0	0.0835	0.2315	0.6935	0.7929	0.9764	1.5882	1.7289	2.0002
the belt t (s)	3-EO	0	0.0836	0.2301	0.6874	0.7865	0.9670	1.5685	1.7063	1.9692
	4-EO	0	0.0841	0.2312	0.6925	0.7910	0.9733	1.5800	1.7158	1.9830

Table 3. Belt test results at a temperature of 7.5 ± 2.0 °C.

Rolling Distance of the Carriage on the Belt s (m)		Probe Tacho 1			Probe Tacho 2			Probe Tacho 3		
		0	0.273	0.772	2.175	2.448	2.947	4.389	4.662	5.161
	1-ST	0	0.0795	0.2243	0.6746	0.7706	0.9487	1.5470	1.6805	1.9417
Mean travel time	1-ST-R1	0	0.0795	0.2248	0.6760	0.7721	0.9508	1.5544	1.6898	1.9564
of the carriage on	2-EO	0	0.0798	0.2250	0.6764	0.7726	0.9509	1.5499	1.6843	1.9460
the belt t (s)	3-EO	0	0.0794	0.2241	0.6739	0.7693	0.9469	1.5426	1.6748	1.9328
	4-EO	0	0.0793	0.2244	0.6737	0.7689	0.9464	1.5375	1.6687	1.9236

Table 4. Belt test results at a temperature of -9.0 ± 2.0 °C.

Rolling Distance of the Carriage on the Belt s (m)		Probe Tacho 1			Probe Tacho 2			Probe Tacho 3		
		0	0.273	0.772	2.175	2.448	2.947	4.389	4.662	5.161
	1-ST	0	0.0830	0.2305	0.6977	0.7990	0.9871	1.6358	1.7891	2.0994
Mean travel time	1-ST-R1	0	0.0802	0.2268	0.6891	0.7881	0.9720	1.6021	1.7486	2.0448
of the carriage on	2-EO	0	0.0802	0.2265	0.6879	0.7862	0.9694	1.5919	1.7351	2.0214
the belt t (s)	3-EO	0	0.0824	0.2291	0.6901	0.7902	0.9747	1.6024	1.7479	2.0399
	4-EO	0	0.0823	0.2288	0.6891	0.7883	0.9719	1.5933	1.7361	2.0206

The data provided in Tables 2–4 were used to calculate the deceleration of the idler set and subsequently to calculate unit indentation rolling resistances. Table 5 contains the values of the idler set deceleration for each investigated temperature, calculated from Equation (4). Subsequently, unit belt indentation rolling resistance values were calculated from Equation (2).

Table 5. Indentation rolling resistance test results at the investigated temperatures.

Belt Symbol	Loading Set Deceleration, a (m/s^2)			Unit Indentation Rolling Resistances, W_e (N/m)		
	23.0 ± 2.0 °C	7.5 ± 2.0	-9.0 ± 2.0	23.0 ± 2.0 °C	7.5 ± 2.0	-9.0 ± 2.0
1-ST	0.8648	0.8974	0.9390	26.10	35.04	46.44
1-ST-R1	0.8712	0.9110	0.9464	27.86	38.76	48.46
2-EO	0.8562	0.8926	0.9184	23.75	33.72	40.79
3-EO	0.8520	0.8904	0.9254	22.60	33.12	42.71
4-EO	0.8416	0.8770	0.9098	19.75	29.45	38.43

Based on the data provided in Table 5, Table 6 was compiled; it contains the indentation rolling resistance test results for different temperatures and for unit loads of 2.3 kN/m. The unit loads correspond to the mean annual loads on belt conveyors in Polish surface lignite mines in Belchatow, Turow, Konin, and Adamow. In other words, they correspond to a situation in which the conveyors are filled with material at 40% capacity, as an average per year. The change in unit rolling resistance is provided in relation to belt 1-ST, which is a standard belt, typically used in the above-mentioned mines. The sign "-" indicates that the belt is characterized by lower indentation rolling resistances in comparison to the standard belt.

Table 6. Test results summary.

Belt Symbol	Mean Unit Indentation Rolling Resistance, N/m	Change of Belt Indentation Rolling Resistances in Relation to Belt 1-ST	
		N/m	%
1-ST	35.86	0	0
1-ST-R1	38.36	2.50	7.0
2-EO	32.75	−3.11	−8.7
3-EO	32.81	−3.05	−8.5
4-EO	29.21	−6.65	−18.5

In comparison to the standard belt, all the tested energy-saving belts show lower rolling resistance values. The energy-saving belts designated with symbols 2-EO and 3-EO have similar mean rolling resistance values at almost 33 N/mm. They are lower by approximately (8.5 ÷ 8.7)% from the value obtained for the standard belt, which was 35.9 kN/m. In comparison to the standard belt, the energy-saving belt 4-EO has an indentation rolling resistance lower by approximately 18.5%. On the other hand, the 1-ST-R1 belt shows indentation rolling resistance higher by 7.0% in comparison to the 1-ST belt. This value is due to the fact that the belt was subjected first to a refurbishment. The costs of new belts incurred by the mines cause the refurbishment of used belts to be profitable. The refurbishment process consists in the removal of damaged covers from the belt and in their replacement with new vulcanized covers made of adequate rubber compound. The refurbishment process also consists in the replacement of damaged cords in the belt core [47]. Out of the total number of steel-cord belts used at the Belchatow mine, almost 40% are refurbished [48].

The tests indicated that it is possible to use steel-cord belts with indentation rolling resistances reduced by 8.5% in comparison to standard belts. The test results for the 4-EO belt suggest a potential to reduce these resistances even further.

In comparison to the 3-EO belt, the 4-EO belt is different in the type and design of the reinforcement plies used in the covers, which may be of significance for the obtained belt indentation rolling resistance values.

The application scale of refurbished belts indicates that the refurbishment should be performed with the use of improved materials allowing lower indentation rolling resistances.

4. Discussion

In order to demonstrate the practical advantages resulting from the use of energy-saving belts, this article also included calculations of the power demand of a conveyor drive mechanism, as measured on a belt conveyor operated at the Belchatow mine. The calculations and the analyses were performed for a standard conveyor transporting overburden. The calculations allowed for the dynamic parameters of belt covers that were identified in separate research works. The calculations were performed with a professional QNK-TT software application, which employs an algorithm for calculating conveyor resistances to motion [49]. The technical data used in the calculations are listed in Table 7, and the operating data in Table 8. The conveyor route consists of three sections having the following lengths: section 1—60 m, section 2—1065 m, and section 3—80 m. The lift height in sections 2 and 3 is 9 m and 6 m, respectively. Section 1 is flat. The vertical curve radius for all sections is 0 m.

Table 7. Conveyor technical data.

Parameter Description	Unit	Value
Belt type	-	ST
Belt strength	kN/m	3150
Belt width	mm	2250
Cord diameter	mm	7.6
Type of upper idler sets	-	Articulated
Upper idler troughing angle	deg	45
Upper idler spacing	m	1.2
Upper idler mass	kg	39.1
Mass of rotating parts in the upper idler	kg	29.5
Upper idler diameter	m	0.19
Upper idler axle length	m	0.8
Side skirt length at the feed point	m	8.5
Type of bottom idlers	-	With rings
Type of bottom idler sets	-	Articulated
Bottom idler spacing	m	6
Number of bottom idlers	pcs	2
Bottom idler troughing angle	deg	10
Bottom idler mass	kg	42.4
Mass of rotating parts in the bottom idler	kg	35.01
Bottom idler diameter	m	0.19
Bottom idler axle length	m	1.3
Belt velocity	m/s	5.98
Idler rotational coefficient a_l	-	3.29
Idler rotational resistance coefficient b_l	-	0.34
Idler–belt friction coefficient	-	0.25
Bottom cover thickness	mm	7
Core thickness	mm	7
Top cover thickness	mm	14
Plate stiffness of the belt	Nm	40.51
Belt damping factor at bending	-	0.58
Linear modulus of belt elasticity	kN/m	225
Elastic modulus of bottom cover	$N/m^2 \cdot 10^6$	40
Elastic modulus of the core	$N/m^2 \cdot 10^6$	25
Mass of 1 m^2 of the belt	kg/m^2	46.91
Belt stiffness	-	0.17
Belt damping factor at indentation	-	0.5
Head and drive pulley diameter	m	1.6
Diameter of the remaining pulleys	m	1.25
Belt–pulley friction coefficient	-	0.35

Table 8. Operating data.

Parameter Description	Unit	Value
Lateral wandering	-	None
Operating conditions	-	Average
Ambient temperature	°C	20
Material angle of repose	deg	15
Material–side skirt friction coefficient	-	0.6
Internal friction angle	deg	35
Bulk density	kg/m^3	1700
Friction angle between the idler and the belt	rad	0.26

The parameters shaded in Tables 7 and 8 were edited during the calculations depending on the applied elements of the conveyor (the idler and the belt), the fill degree with material, which was translated into vertical forces acting on the three-idler set, and the ambient temperature.

The power of the conveyor drive mechanism was calculated depending on the ambient temperature. The assumed temperature range was from $-25\,°C$ to $30\,°C$. The results of the calculations are shown in Table 9.

Table 9. Calculation results for the conveyor drive power.

Temperature, (°C)	Drive Power, (kW)				
	Belt Symbol				
	1-ST	1-ST-R1	2-EO	3-EO	4-EO
−25	2361	2580	2482	2433	2450
−20	2177	2360	2270	2224	2229
−10	1866	2026	1943	1908	1860
0	1673	1800	1715	1696	1483
10	1552	1660	1553	1558	1288
20	1484	1580	1450	1476	1239
30	1457	1556	1393	1440	1231

The calculation results indicate that the power demand of the conveyor drive mechanism increases together with the decrease of ambient temperature. This phenomenon is directly caused by the fact that a decrease in temperature results in an increase of rolling resistances. Conveyor drive mechanisms show the most limited power demand at positive temperatures. Positive temperatures also cause the greatest fluctuations of the power demand. In negative temperatures, the power demand is similar for all the belts. Below $-10\,°C$, the standard belt shows the lowest power demand, which suggests that its energy-efficiency is the highest at such temperatures. This fact suggests, on the other hand, that a belt that is energy-saving at positive temperatures, may not necessarily be energy-saving at negative temperatures. Belts 2-EO, 3-EO, and 4-EO are examples of this tendency. It may be an important observation for both users and manufacturers of conveyor belts, as it suggests that belts should be selected with consideration paid to their future operating temperatures. For example, belt 4-EO will prove to be energy-saving if it is operated in a climate in which temperatures rarely drop to approximately $0\,°C$. In the case of belt 1-ST with a standard bottom cover, the increases in power demand are lower than in the case of the 4-EO belt, with the bottom cover having the lowest rolling resistances. This fact means that the bottom cover of a standard belt is less sensitive to ambient temperature changes. At extreme temperatures, the increase in the power demand of the conveyor drive mechanism may be almost two-fold.

Figure 5 shows a comparison of the power demand from a conveyor drive mechanism for the 1-ST standard belt used in lignite mines and for the 4-EO energy-saving belt. The power of the conveyor drive mechanism is shown against the mean ambient temperature in successive months of the year. The blue line indicates mean monthly temperatures in the area of the Belchatow mine. The bottom cover of the 1-ST belt is made of a rubber compound commonly used by the manufacturer of this type of belts. The bottom cover of the 4-EO belt has lower rolling resistance values in comparison to the standard 1-ST belt.

Figure 6 shows a comparison of the power demand from a conveyor drive mechanism for the 1-ST standard belt and for the 1-ST-R1 refurbished belt used in lignite mines. The bottom cover of this belt is manufactured from a rubber compound which has the highest rolling resistance values of all the tested belts.

In comparison to the standard belt, the refurbished belt generates, on average per annum, a power demand higher by approximately 4.8%.

By installing a steel-cord conveyor belt with the lowest indentation rolling resistance and the highest energy-efficiency, in place of the standard belt used to date in the mine, the power consumption of the conveyor drive mechanism can be reduced by approximately 15.3% on average per annum. If this belt is used in place of the refurbished belt, the power demand is reduced by more than 20%.

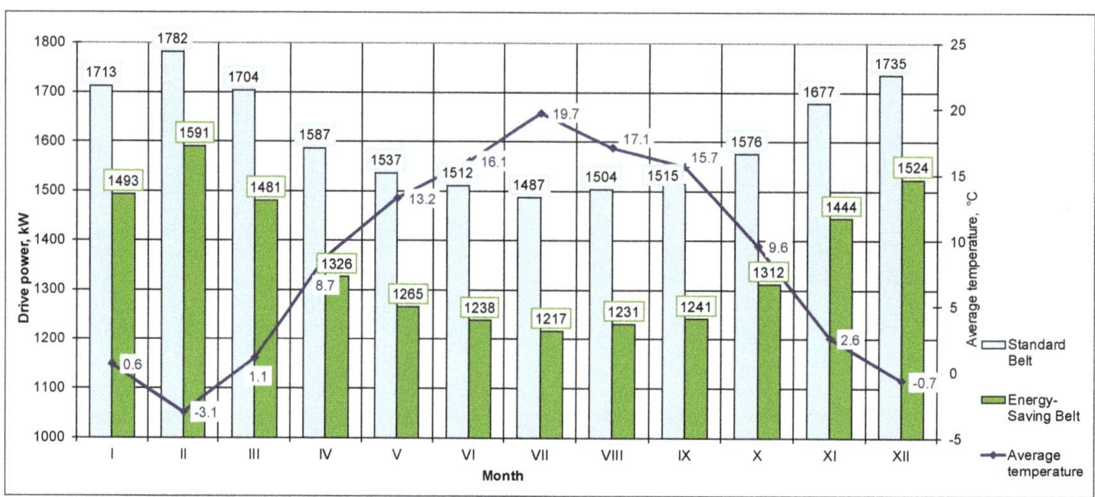

Figure 5. Relationship between power consumption and mean monthly temperatures for the 1-ST standard belt and for the 4-EO energy-saving belt.

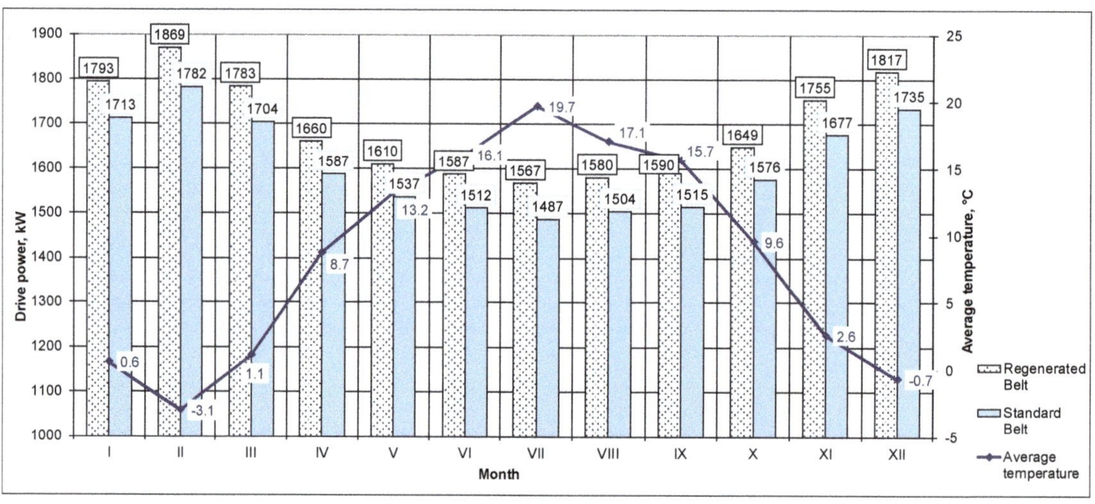

Figure 6. Relationship between power consumption and mean monthly temperatures for the 1-ST standard belt and for the 1-ST-R1 refurbished belt.

Although it generates lower belt indentation rolling resistances, the energy-saving belt is more sensitive to ambient temperatures in comparison to the standard belt and to the refurbished belt. The power consumption difference within the range of extreme temperatures observed in February and in July was 374 kW for the energy-saving belt, 295 kW for the standard belt and 302 kW for the refurbished belt.

The proposed method for measuring belt indentation rolling resistance on the "inclined plane" test rig is the result of long-lasting research into belt conveyor resistance to motion, performed at Wroclaw University of Science and Technology [50,51]. Acquiring proper dimensions of the conveyor drive mechanism is a very important design task. It can be performed based on conveyor resistances to motion. These resistances can be most advantageously estimated from unit resistance methods, which were developed at different research centers [52]. The improvement of the calculation methods to verify the identified relationships requires laboratory tests. This article presented an experimental test method for measuring the greatest component of conveyor belt motion resistances, i.e., indentation rolling resistances. This method consisted in measuring a carriage with two idlers rolling on the conveyor belt. The obtained results may serve not only to verify the calculation methods, but also to compare the rolling resistances of different belt types and designs. The identification of the impact of various factors on belt indentation rolling resistances is also a basis for any activities aimed at finding solutions to help reduce the electric energy consumption of belt conveyor drive mechanisms.

5. Conclusions

The "inclined plane" test rig developed for the purpose of identifying belt indentation rolling resistances allowed quick and easy measurements. Tests performed with the use of this "inclined plane" may help research optimal parameters for rubber compounds and optimal design solutions for conveyor belt bottom covers.

A complex analysis of the influence of the bottom cover design and of the rubber compounds used in bottom covers on indentation rolling resistances must include a number of criteria, such as the sensitivity of rubber compounds to temperature changes. The operating temperature of a particular conveyor thus becomes an important factor.

The power consumption of conveyor drive mechanisms is possible if the rubber used in the bottom cover of a steel-cord belt has appropriately selected mechanical parameters that reduce indentation rolling resistances. At a mean annual temperature of 8.3 °C, if a standard belt used to date in Polish lignite mines is replaced with an energy-saving belt with a bottom cover generating lower indentation rolling resistances, the power consumption of the conveyor main drive mechanism can be reduced by approximately 15%.

Refurbished conveyor belts should be withdrawn from service. They generate higher rolling resistances and thus increase the power consumption of conveyor drive mechanisms. The application scale of refurbished belts at the Belchatow mine, where they account for 40% of all operated steel-cord belts, causes their operating cost to be very high.

Author Contributions: Conceptualization, M.B.; methodology, M.B.; software, M.B.; validation, M.B. and M.H.; formal analysis, M.H.; investigation, M.B.; resources, M.B.; data curation, M.H.; writing—original draft preparation, M.B.; writing—review and editing, M.H.; visualization, M.B.; supervision, M.H.; project administration, M.H.; funding acquisition, M.H. Both authors have read and agreed to the published version of the manuscript.

Funding: The research work was cofounded with the research subsidy from the Polish Ministry of Science and Higher Education granted for 2021.

Institutional Review Board Statement: Not applicable.

Informed Consent Statement: Not applicable.

Data Availability Statement: The data presented in this study are available on request from the corresponding author.

Acknowledgments: The authors would like to thank the Belchatow mine for providing belt specimens for tests.

Conflicts of Interest: The authors declare no conflict of interest.

References

1. Yao, Y.; Zhang, B. Influence of the elastic modulus of a conveyor belt on the power allocation of multi-drive conveyors. *PLoS ONE* **2020**, *15*, e0235768. [CrossRef]
2. Gallagher, D. Low rolling resistance for conveyor belts. In *International Rubber Conference*; The Goodyear Tire & Rubber Company Publishing: Melbourne, Australia, 2000; Volume 10, pp. 245–261.
3. Munzenberger, P.; Wheeler, C. Laboratory measurement of the indentation rolling resistance of conveyor belts. *Measurement* **2016**, *94*, 909–918. [CrossRef]

4. Bajda, M. The Impact of the Conveyor Belt Rubber Cover on Its Rolling Resistance over Idlers. Ph.D. Thesis, Wroclaw University of Technology, Wrocław, Poland, 2009. Not published. (In Polish)

5. Konieczna-Fuławka, M. Method for Determining Belt Indentation Rolling Resistance. Ph.D. Thesis, Wroclaw University of Science and Technology, Wroclaw, Poland, 2019. Not published. (In Polish)

6. Gładysiewicz, L.; Konieczna-Fuławka, M. Influence of idler set load distribution on belt rolling resistance. *Arch. Min. Sci.* **2019**, *64*, 251–259.

7. Suchorab, N. Specific energy consumption—The comparison of belt conveyors. *Min. Sci.* **2019**, *26*, 263–274. [CrossRef]

8. Zang, S.; Tang, Y. Optimal scheduling of belt conveyor systems for energy efficiency—With application in a coal-fired power plant. In Proceedings of the 2011 Chinese Control and Decision Conference (CCDC), Mianyang, China, 23–25 May 2011; pp. 1434–1439. [CrossRef]

9. Mathaba, T.; Xia, X. Optimal and energy efficient operation of conveyor belt systems with downhill conveyors. *Energy Effic.* **2017**, *10*, 405–417. [CrossRef]

10. Akparibo, A.R.; Normanyo, E. Application of resistance energy model to optimising electric power consumption of a belt conveyor system. *Int. J. Electr. Comput. Eng.* **2020**, *10*, 2861–2873. [CrossRef]

11. Mathaba, T.; Xia, X. A parametric energy model for energy management of long belt conveyors. *Energies* **2015**, *8*, 13590–13608. [CrossRef]

12. Hager, M.; Hintz, A. The Energy Saving-Design of Belts for Long Conveyor Systems. *Bulk Solids Handl.* **1993**, *13*, 749–758.

13. Kawalec, W. Long distance belt conveyor systems for transportation lignite. *Transp. Przemysłowy I Masz. Rob.* **2009**, *1*, 2–9. (In Polish)

14. Zimroz, R.; Hardygóra, M.; Blazej, R. Maintenance of Belt Conveyor Systems in Poland—An Overview. In *Proceedings of the 12th International Symposium Continuous Surface Mining—Aachen 2014*; Lecture Notes in Production Engineering; Niemann-Delius, C., Ed.; Springer: Berlin/Heidelberg, Germany, 2014. [CrossRef]

15. Gładysiewicz, L.; Hardygóra, M.; Kawalec, W. Belt conveying in the Polish mining industry—Selected topics. *Bulk Solids Handl.* **2004**, *24*, 1–7.

16. Bajda, M.; Jurdziak, L.; Konieczka, Z. Comparison of electricity consumption by belt conveyors in a brown coal mine. Part 2. Study of the belt conveyors capacity influence. *Górnictwo Odkryw.* **2019**, *3*, 26–38. (In Polish)

17. Bajda, M.; Jurdziak, L.; Konieczka, Z. Comparison of electricity consumption by belt conveyors in a brown coal mine. Part 3. Correction of the forecasts of the unit energy consumption by temperature influence. *Górnictwo Odkryw.* **2019**, *4*, 12–23. (In Polish)

18. Krawiec, P.; Warguła, L.; Małozięć, D.; Kaczmarzyk, P.; Dziechciarz, A.; Czarnecka-Komorowska, D. The Toxicological Testing and Thermal Decomposition of Drive and Transport Belts Made of Thermoplastic Multilayer Polymer Materials. *Polymers* **2020**, *12*, 2232. [CrossRef] [PubMed]

19. Krawiec, P.; Warguła, L.; Czarnecka-Komorowska, D.; Janik, P.; Dziechciarz, A.; Kaczmarzyk, P. Chemical compounds released by combustion of polymer composites flat belts. *Sci. Rep.* **2021**, *11*, 8269. [CrossRef]

20. Gładysiewicz, L. *Belt Conveyors. Theory and Calculations*; Wroclaw University of Technology Press: Wrocław, Poland, 2003. (In Polish)

21. Lechmann, H.P. Der Walkwiederstand von Gummigurtförderern. *Forsch. Geb. Ing. A* **1954**, *20*, 97–107. [CrossRef]

22. Oehmen, H.H. Stetigförderer. Vorlesungsmanuskript am Lehrstuhl und Institute für Fordertechnik. Dissertation Thesis, Hanover University, Hanover, Germany, 1969.

23. Kehlert, H. Untersuchungen zur Bestimmung des Bewegungswiederstandes von Gurtbandfordereren. Dissertation Thesis, Magdeburg University, Hanover, Germany, 1977.

24. Jonkers, C.O. The indentation rolling resistance of belt conveyors. *Ford. Heb.* **1980**, *30*, 312–318.

25. Spaans, C. *The Indentation Resistance of Belt Conveyors*; Technical Report WTHD 103; Laboratory for Transport Engineering: Delft, The Netherlands, 1978.

26. Kostrzewa, H. Einfluss der Fordergurttemperatur auf dem Eindruckrollwiederstand zwischen Fordergurt und Tragrolle. *Ford. Heb.* **1985**, *35*, 840–842.

27. Limberg, H. Untersuchungen der Trummbezogenen Bewegungswiederstande von Gurtforderanlagen. Dissertation Thesis, Hanover University, Hanover, Germany, 1988.

28. Hintz, A. Einfluss des Gurtaufbase auf dem Energieverbrauch von Gurtforderanlagen. Dissertation Thesis, Hanover University, Hanover, Germany, 1993.

29. Wheeler, C. Analysis of the Main Resistance of Belt Conveyors. Ph.D. Thesis, University of Newcastle, Newcastle, Australia, 2003.

30. Hötte, S.; von Daacke, S.; Schulz, L.; Overmeyer, L.; Wennekamp, T. The Way to the DIN 22123—Indentation Rolling Resistance of Conveyor Belts. *BulkSolids Handl.* **2012**, *6*, 48–52.

31. Australian Standard AS 1334.13:2017. Methods of Testing Conveyor and Elevator Belting Determination of Indentation Rolling Resistance of Conveyor Belting. Available online: https://infostore.saiglobal.com/en-au/standards/AS-1334-13-2017-99439_SAIG_AS_AS_209059/ (accessed on 20 September 2021).

32. German Standard DIN 22123:2012-10. Conveyor Belts—Indentation Rolling Resistance of Conveyor Belts Related to Belt Width—Requirements, Testing. Available online: https://www.iso.org/standard/79942.html (accessed on 20 September 2021).

33. Wozniak, D. Laboratory tests of indentation rolling resistance of conveyor belts. *Meas. Lond.* **2020**, *150*, 107065. [CrossRef]

34. He, D.; Pang, Y.; Lodewijks, G. Green operations of belt conveyors by means of speed control. *Appl. Energy* **2017**, *188*, 330–341. [CrossRef]
35. Markos, P.; Dentsoras, A. An index that correlates service performance and energy consumption of belt conveyors. *FME Trans.* **2018**, *46*, 313–319. [CrossRef]
36. Sarathbabu Goriparti, N.V.; Murthy, C.S.N.; Aruna, M. Prediction of energy efficiency of main transportation system used in Underground Coal Mines—A Statistical Approach. In *International Conference on Emerging Trends in Engineering (ICETE)*; Springer: Cham, Switzerland, 2020; pp. 337–344.
37. Köken, E.; Lawal, A.I.; Onifade, M.; Özarslan, A. A comparative study on power calculation methods for conveyor belts in mining industry. *Int. J. Mining. Reclam. Environ.* **2021**. [CrossRef]
38. Mu, Y.; Yao, T.; Jia, H.; Yu, X.; Zhao, B.; Zhang, X.; Ni, C.; Du, L. Optimal scheduling method for belt conveyor system in coal mine considering silo virtual energy storage. *Appl. Energy* **2020**, *275*, 115368. [CrossRef]
39. Kawalec, W.; Suchorab, N.; Konieczna-Fuławka, M.; Król, R. Specific Energy Consumption of a Belt Conveyor System in a Continuous Surface Mine. *Energies* **2020**, *13*, 5214. [CrossRef]
40. Kawalec, W.; Król, R. Generating of electric energy by a declined overburden conveyor in a continuous surface mine. *Energies* **2021**, *14*, 4030. [CrossRef]
41. Jurdziak, L.; Błażej, R.; Bajda, M. Conveyor Belt 4.0. In *Intelligent Systems in Production Engineering and Maintenance. ISPEM 2018. Advances in Intelligent Systems and Computing*; Burduk, A., Chlebus, E., Nowakowski, T., Tubis, A., Eds.; Springer: Cham, Switzerland, 2018; Volume 835. [CrossRef]
42. Warguła, Ł.; Kukla, M.; Lijewski, P.; Dobrzyński, M.; Markiewicz, F. Impact of Compressed Natural Gas (CNG) Fuel Systems in Small Engine Wood Chippers on Exhaust Emissions and Fuel Consumption. *Energies* **2020**, *13*, 6709. [CrossRef]
43. Ji, J.; Miao, C.; Li, X. Research on the energy-saving control strategy of a belt conveyor with variable belt speed based on the material flow rate. *PLoS ONE* **2020**, *15*, e0227992. [CrossRef]
44. Kulinowski, P.; Kasza, P.; Zarzycki, J. Influence of design parameters of idler bearing units on the energy consumption of a belt conveyor. *Sustainability* **2021**, *13*, 437. [CrossRef]
45. Gładysiewicz, L.; Król, R.; Kisielewski, W. Measurements of loads on belt conveyor idlers operated in real conditions. *Measurement* **2019**, *134*, 336–344. [CrossRef]
46. Kasztelewicz, Z. *Metoda Programowania Zagospodarowania Złóż W Wieloodkrywkowej Kopalni Węgla Brunatnego*; Uczelniane Wydawnictwa Naukowo-Dydaktyczne Akademii Górniczo-Hutniczej im: Stanisława, Poland, 2005. (In Polish)
47. Błażej, R.; Jurdziak, L.; Kirjanów-Błażej, A.; Kozłowski, T. Identification of damage development in the core of steel cord belts with the diagnostic system. *Sci. Rep.* **2021**, *11*, 12349. [CrossRef]
48. Wlazłowski, W.G. Ocena Technologii Napraw Taśm Przenośnikowych Stosowanych w Kopalni Bełchatów. Dissertation Thesis, Wroclaw University of Science and Technology, Wrocław, Poland, 2021. Not published. (In Polish)
49. Kulinowski, P.; Kawalec, W. Program Komputerowy QNK-TT do Wspomagania Projektowania Przenośników Taśmowych. (In Polish). Available online: http://www.EnterTECH.com.pl/QNK (accessed on 29 July 2021).
50. Gładysiewicz, L.; Król, R.; Kisielewski, W.; Kaszuba, D. Experimental determination of belt conveyors artificial friction coefficient. *Acta Montan. Slovaca* **2017**, *22*, 206–214.
51. Gładysiewicz, L.; Król, R.; Bukowski, J. Tests of belt conveyor resistance to motion. *Eksploat. I Niezawodn. Maint. Reliab.* **2011**, *3*, 17–25.
52. Geesmann, F. Experimentelle and theoretischeuntersuchungen der Bewegungswiderstände von Gurtförderanlagen (Experimental and Theoretical Investigations of the Motion Resistances of Belt Conveyor Systems). Ph.D. Thesis, Hanover University, Hanover, Germany, 2001.

Article

Aspects of Selecting Appropriate Conveyor Belt Strength

Dariusz Woźniak * and Monika Hardygóra

Faculty of Geoengineering, Mining and Geology, Wroclaw University of Science and Technology, Na Grobli 15 St., 50-421 Wrocław, Poland; monika.hardygora@pwr.edu.pl
* Correspondence: dariusz.wozniak@pwr.edu.pl

Abstract: Breaks in the so-called "continuous" (unspliced) belt sections, and not in the spliced areas, are infrequent but do happen in practice. This article presents some aspects, which may account for such breaks in conveyor belts. It indicates the so-called "sensitive points" in design, especially in the transition section of the conveyor belt and in identifying the actual strength of the belt. The presented results include the influence of the width of a belt specimen on the identified belt tensile strength. An increase in the specimen width entails a decrease in the belt strength. The research involved develops a universal theoretical model of the belt on a transition section of a troughed conveyor in which, in the case of steel-cord belts, the belt is composed of cords and layers of rubber, and in the case of a textile belt, of narrow strips. The article also describes geometrical forces in the transition section of the belt and an illustrative analysis of loads acting on the belt. Attention was also devoted to the influence of the belt type on the non-uniform character of loads in the transition section of the conveyor. A replacement of a conveyor belt with a belt having different elastic properties may increase the non-uniformity of belt loads in the transition section of the conveyor, even by 100%.

Keywords: conveyor belt; belt tensile strength; transition section; theoretical model

Citation: Woźniak, D.; Hardygóra, M. Aspects of Selecting Appropriate Conveyor Belt Strength. *Energies* **2021**, *14*, 6018. https://doi.org/10.3390/en14196018

Academic Editor: Nikolaos Koukouzas

Received: 31 July 2021
Accepted: 17 September 2021
Published: 22 September 2021

Publisher's Note: MDPI stays neutral with regard to jurisdictional claims in published maps and institutional affiliations.

1. Introduction

Belt conveyors are an indispensable means of transport in a number of industry branches–most importantly, in the extraction industry, in smelting and coking plants, and in power plants, but also in the chemical industry, in civil engineering and in agriculture. Belt conveyors are also vital in shipping ports, where they are used to transport, load and unload bulk materials. In the extraction industry, conveyor belt transportation systems are used both in surface and in underground mining. The belt is the most expensive part of the belt conveyor. Its purchase cost represents 50–60% of the cost of the entire conveyor. The belt is also the least durable element of the conveyor. The belt is, thus, an element crucial for the effective and reliable operation of the conveyor and significantly influences the transportation costs [1].

The selection of a conveyor belt capable of performing a particular transportation task when installed on a belt conveyor is primarily informed by the tensile strength of this belt. It is intended to ensure that the forces in the operating belt will not lead to it breaking, i.e., to an event that is dangerous to both the personnel and to the conveyor. Typically, in the belt conveyor design process, the values of safety factors are empirically identified [2,3].

The belt is installed on the conveyor in a closed loop. The type of the splices used in the belt depends mainly on the design of the belt core. In most cases, splices have the lowest strength in the entire belt. Nevertheless, occasionally, breaks develop not in the area of the splice, but in the so-called "continuous" belt section. Reasons for such events are an object of investigation. This article analyzes two aspects, which deserve attention when selecting an appropriate belt to match the conveyor—the non-uniformity of belt loads in the transition section of the conveyor, where the belt changes its shape from troughed into flat to enter the pulley, and the influence of the specimen width on the belt strength. Aspects related to belt damage during its operation (punctures, cuts, etc.) are here omitted, although obviously, they are important and monitored by conveyor operators [4–7]. The analysis here presented is supported by many years of experience gained as a result of testing conveyor belts at the Belt Conveying Laboratory, Wroclaw University of Science and Technology (WUST).

When implementing a particular transportation task, the force levels in the belt and the power of the drive system are identified after calculating conveyor resistances to motion and the minimal tensile force required to ensure frictive engagement between the pulley and the belt. Köken et al. reviewed and compared leading methods used in belt conveyor calculations [8].

In the transition section of the conveyor, the outermost load-carrying elements of the core accept additional tensile loads, while the loads in the central part of the belt decrease. In the existing calculation methods, the length of the transition section should be defined in such a manner that the non-uniformity of belt loads is reduced and unit forces in the belt are not exceeded. That CEMA method [9] defines the minimum length of the transition section depending on the geometrical parameters of the section and differentiates between only two groups of belts: with steel cord and with textile core. Importantly, however, textile core belts may significantly differ with respect to their elastic properties and show different reactions in response to geometry changes of the transition section. The Fenner–Dunlop method has a similar approach to the identification of the transition section [10]. In the DIN method [11], the length of the transition section is calculated on the basis of simplified relationships, but with allowance for the longitudinal elastic modulus of the belt.

Accurate identification of the stress state in the belt along the transition section requires a model that allows for not only the longitudinal elastic modulus of the belt, but also for the interactions between the adjacent cables or straps in the belt. Oehmen [12] and Hager and Tappeiner [13] focused in their research on identifying the strain state in the belt along the transition section. Schmandra [14] presented a general theoretical model of a steel cord belt, allowing for the interactions between the adjacent cables. The literature also mentions implementations of the finite element method in the modeling of a belt along the transition section of the conveyor [15–18]. The analysis in [17] focused on the influence of the elastic modulus in the longitudinal and transverse directions of the belt-on-belt load non-uniformity.

The research here presented involves developing a universal theoretical model of the belt along a transition section of a conveyor in which, in the case of steel-cord belts, the belt is composed of cords and layers of rubber, and in the case of a textile belt, of narrow strips. An analysis was performed into how the non-uniformity of the belt load along the transition section of the conveyor will change if belts with different cores are used. The research also involved tests of the influence of the specimen width on the belt tensile strength. The literature does not mention any results of similar previous studies.

2. Materials and Methods

2.1. Selection of Belt Strength

In the design phase of a belt conveyor, the selection of a conveyor belt typically consists of selecting the material and core design, and also in deciding on the thickness and the material of the protective covers. An optimal design of the belt core is achieved on the basis of a detailed technical and economic analysis, which allows for such factors as follows [1]:

- Belt nominal strength in relation to the forces in the belt on an operating conveyor;
- Conveyor length;
- Potential for splicing belt sections;
- Fatigue life of splices;
- Loads in the material feeding points;
- Belt operating conditions, fire and explosion protection;
- Properties of the transported material (size of the lumps, their angularity, temperature, etc.).

Nominal strength of a textile-core belt is identified, according to the following relationship [3]:

$$K_n \geq \frac{S_{r\,max}}{1000 * B} \cdot k_e \cdot k_b \tag{1}$$

where:

K_n—nominal strength, kN/m or N/mm;
B—belt width, m;
$S_{r\,max}$—maximum strength in the belt during start-up, N;
k_e—safety factor allowing for the operating conditions;
k_b—safety factor allowing for lower strength in the splice.

With the values of individual factors assumed in accordance with the recommendations found in the literature [2,3,11], the total safety factor in relationship to the forces in steady motion for textile belts will remain within the range of 9 ÷ 12. In terms of design standards, this level of safety is high. When designing long, high-capacity conveyors with steel-cord belts, proposals are made to lower this safety standard or to use the so-called operational belt safety factor [19]. In the design phase, however, note should be made that as the belt moves on the conveyor, it is subjected to such states as those induced by horizontal and vertical curves or transition sections, which entail significant momentary variations of specific forces in the belt cross-section. This phenomenon is related to the fact that individual threads in the warp of the belt core move along different paths, due to the geometrical forces existing in the belt.

The above applies particularly to the transition section of a conveyor in which the belt experiences the greatest forces, as is the case mostly in the section of belt transition from a troughed shape into a flat shape, at a point in which the belt enters the drive pulley. The difference between the maximum and the minimum specific force in the belt cross-section on the pulley depends on the geometry of the system and on the dynamic modulus of the belt [12–17]. For long conveyors and for two-parameter belt models, this modulus is assumed to be equal to the longitudinal modulus of elasticity. Therefore, designs of the geometry of the transition section are based on the knowledge of the elastic properties of the belt operated on the conveyor. Unfortunately, belt manufacturers rarely include the longitudinal modulus of elasticity in their product specifications. This modulus is identified in laboratory tests, according to the method described in ISO 9856. Research works performed in the Belt Conveying Laboratory at WUST include identifying elastic properties of conveyor belts with cores of different designs. Table 1 presents illustrative values of the elastic modulus for conveyor belts.

Table 1. Results of elastic modulus tests for conveyor belts.

Belt Type	Longitudinal Modulus of Elasticity E, N/m
multi-ply polyamide PP 1400/4 ÷ 2000/4	$8 \times 10^6 \div 12 \times 10^6$
multi-ply polyester-polyamide EP 800/4 ÷ 2000/4	$8 \times 10^6 \div 20 \times 10^6$
solid woven PWG EP(B)PB 1000/1 sw ÷ 2500/1 sw	$15 \times 10^6 \div 28 \times 10^6$
aramid single-ply DP 2000/1 ÷ 3150/1	$68 \times 10^6 \div 76 \times 10^6$
steel-cord ST 2000	116×10^6

Longitudinal modulus of elasticity depends on the strength of the belt, but most importantly on the fabric in the carcass, its weave and core structure. Belts of a similar strength type may have significantly different elastic moduli, e.g., the modulus in a multi-ply polyamide belt with plain-woven plies is two-fold lower than in a belt with a solid-woven carcass and five-fold lower than in an aramid belt with a straight warp carcass. Steel-cord belts have the highest elastic modulus. A higher modulus in the belt with identical geometrical forces in the transition section will mean a greater variety of specific forces in the belt cross-section. Replacement of a conveyor belt with a belt having a different design but identical geometrical parameters of the transition section may, thus, lead to exceeding the limit forces in the belt. This issue was analyzed with the use of a dedicated theoretical model of the belt in the transition section of the conveyor (Section 3).

2.2. Verification of Actual Belt Strength

Conveyor belt strength is identified in laboratory pull tests, in accordance with the method described in ISO 283. This test is commonly performed as part of belt quality control procedures in laboratories at manufacturing plants. Due to technical limitations of the test equipment, the belt strength is identified by testing specimens having a width significantly smaller than the width of the entire belt installed on the conveyor. This fact leads to a question of whether the strength tested on a small specimen corresponds to the strength of the belt having a full width. For this reason, tests were performed into how the test scale influences the test result. The tests at the Belt Conveying Laboratory, WUST, involved identifying tensile strength for the same belt, but for specimens having different widths in the reduced area. Due to the technical capabilities of the test machine, the maximum width of the tested belt specimen was 50 mm. All samples were paddle-shaped and prepared with the same radius R = 560 mm—only the width in the reduced area changed. Five different specimen widths were used: 20, 25, 30, 40 and 50 mm. The tests were performed for a 3-ply belt with polyamide carcass (belt A) and for a 4-ply belt with a polyester–polyamide carcass (belt B). The results of the tests are shown in Figure 1. In the case of both belts, a decrease in the tensile strength was observed to accompany an increase in the specimen width. Whether this phenomenon results from the test method— the specimen shape, change in its width to curve radius ratio, and change in its width to length ratio—or from the strength characteristics of such multi-layer and multi-fiber structures as belts, required further investigations. The tests indicated that an increase in the specimen width entails a decrease in its elongation at break (Figure 2); this fact can be attributed to the strength characteristics of textile belts. If the strength tests are performed for a single thread, the strength of the entire structure comprising such threads is not equal to the multiple of the single thread. This is because each thread in such a structure may have a different preliminary load due to, for example, the preliminary tension, weave, or thermal shrinkage that occurs during the belt production process. In such a case, the break is a gradual process: the threads with the highest preload will break first, followed by the remaining threads. As a consequence, the strength will be lower than the multiple of the strength of a single thread.

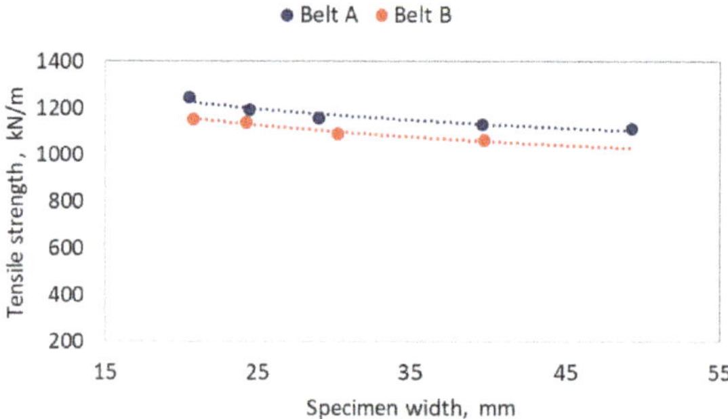

Figure 1. Influence of belt width on the results of belt strength tests.

The tests demonstrated changes in the belt tensile strength, which depend on the specimen width and which can be described with a quadratic function. If assumed that the changes result solely from the strength characteristics of such structures as a conveyor belt, and not from the test method or from other reasons, and if also assumed that the trend identified in the test is constant, then the actual strength of a 1200 mm wide belt may be estimated at 60% of the strength measured in standardized tests. As the safety factor used in the process of selecting belt strengths is relatively high (above 9), the above estimation seems appropriate. This aspect is nevertheless important when using belts of greater widths, or when designing curve geometries along the route and the conveyor transition section, especially in the zone of the highest specific forces in the belt.

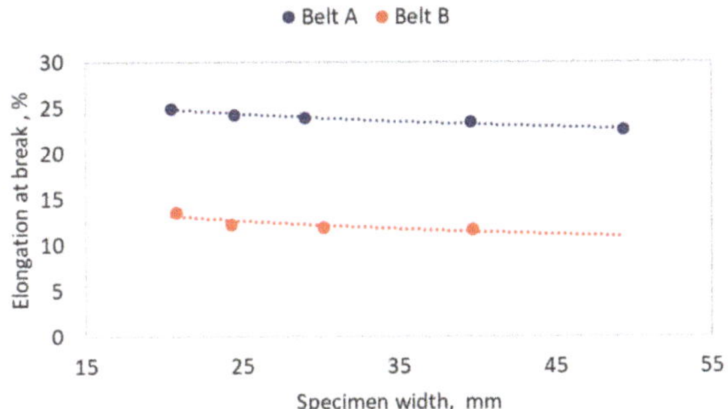

Figure 2. Influence of belt width on the results of belt elongation at break tests.

3. Theoretical Model of the Belt on the Transition Section of the Conveyor

3.1. Preliminary Assumptions behind the Model

1. The belt is considered as a multi-strip flat tension member, globally carrying only longitudinal loads. In the case of a steel-cord belt, the strip is a steel cord with the surrounding rubber, and in the case of a textile belt, the strip is composed of a number of warp threads.
2. The strips are modeled as linear elastic elements.
3. The rubber that connects the adjacent cords or weft threads is responsible for the mutual interaction of the adjacent strips, which is modeled with the use of the artificial modulus of non-dilatational strain.
4. The longitudinal rigidity of the strips and the artificial modulus of non-dilatational strain between the strips are determined on the basis of laboratory tests.
5. The considered belt section has a length $L_e + K$, where L_e is the length of the transition section, and K is the length of the section of influence, in which the stresses equalize.
6. The belt is subjected to tension at a constant force F_N.
7. No external loads are observed along the length L_e of the transition section (i.e., motion resistances are absent, and the impact of gravity is negligible), which means that in each cross-section, the sum of forces in the strips is equal to the tensile force F_N. The influence of the pulley is also negligible.
8. The rubber between the cables is not deformed in the lateral direction, and therefore, the distances between the cable axes (the cable pitch) are constant. This assumption allows the equation to be solved as a system of coplanar forces.
9. Calculations are performed for a half of the belt, from the edge to its axis of symmetry (the system is assumed to be symmetrical).

3.2. Geometrical Forces in the Transition Section of the Conveyor

The basis for a model thus formulated is to determine elongations for individual strips. These result from the path traveled by each strip when the belt transitions from a troughed cross-section into a flat cross-section on the pulley. Consideration is paid to the path traveled by steel cords (in the case of steel-cord belts) or to the path of the longitudinal axes of the analyzed strips (in the case of textile belts). The issue was analytically described with the use of the coordinate system shown in Figure 3.

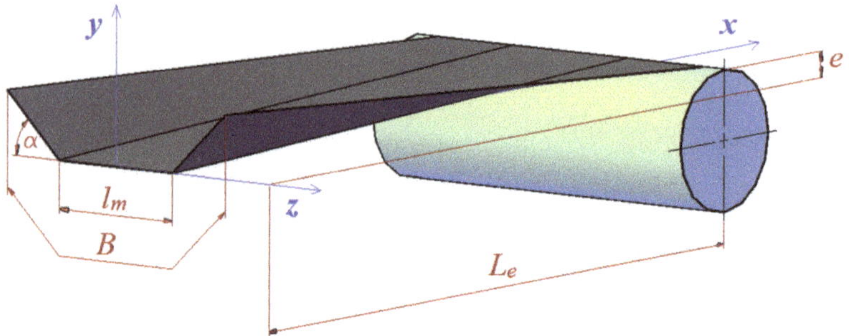

Figure 3. Transition section of the belt conveyor (L_e—length of the transition section; B—belt width; α—trough angle; l_m—length of the central part of the belt trough; e—height difference between the contour of the pulley coat and the central part of the trough).

The length of each cable Δl (each longitudinal axis of the strip) is determined as the Euclidean distance between the position of the cable at the beginning of the coordinate system ($x = 0$) and the position of the cable at the end of the transition section ($x = L_e$).

$$\Delta l = \sqrt{\Delta x^2 + \Delta y^2 + \Delta z^2} \tag{2}$$

Cable elongation Δu in the transition section is the following:

$$\Delta u = \Delta l - L_e \tag{3}$$

At the beginning of the transition section, the belt cross-section is troughed, and the coordinate $x = 0$. The positions of individual cables may be described with the following relationships:

- For $z \leq \frac{1}{2} l_m$

$$z_i = i \cdot t \tag{4}$$

$$y_i = 0 \tag{5}$$

- For $z > \frac{1}{2} l_m$

$$z_i = \frac{1}{2} l_m + \left(i \cdot t - \frac{1}{2} l_m \right) \cos \alpha \tag{6}$$

$$y_i = \left(z_i - \frac{1}{2} l_m \right) \tan \alpha \tag{7}$$

where:
i—cable number $i = 1 \div n$;
n—number of cables across the half of the belt width;
l_m—length of the central part of the belt trough, mm;
t—cable pitch (strip width), mm;
α—trough angle.
At the end of the transition section, the belt cross-section is flat, and the coordinate $x = L_e$. The positions of individual cables may be described with the following relationships:

$$z_i = i \cdot t \tag{8}$$

$$y_i = e \tag{9}$$

where:

e—height difference between the contour of the pulley coat and the central part of the trough, mm.

The belt displacement area determined from the model based only on the geometrical relationships (Figure 4) is a certain simplification of the actual condition, but it is sufficient to analyze the behavior of various belt core structures along the transition section of the conveyor.

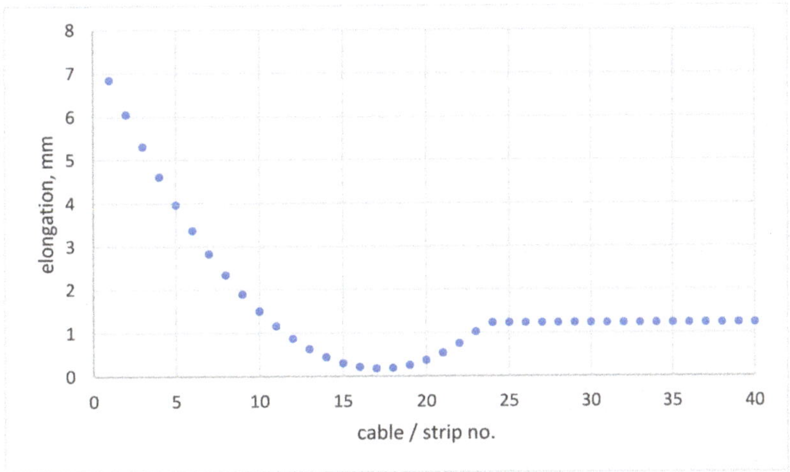

Figure 4. Cable/strip elongation in the transition section of the belt conveyor identified from the geometrical model (L_e = 2.6 m, B = 1.2 m, e = 80 mm, α = 45°).

3.3. Theoretical Basis for the Belt Model in the Transition Section

The model reflects half of the belt, having a length equal to the length of the transition section L_e and the length of the section of influence K. The symmetrical half of the belt core is composed of n cables. The initial load state in the belt core is due to the F_N force in the belt. The cables form a parallel structure, i.e., each cable in the core will be elongated by an identical initial distance u_0. Thus, the initial force in the ith cord F_{oi} is the following:

$$F_{oi} = \frac{u_0}{l}(EA)_i \tag{10}$$

from the condition of the equilibrium of forces as follows:

$$F_N = 2\sum_{i=1}^{n} F_{oi} \tag{11}$$

where:

$(EA)_i$—longitudinal rigidity of the ith cable/strip, N;
l—length of the analyzed belt section, mm.

On the other hand, the forces from the geometrical input related to the transition section of the conveyor in which cable elongations have a non-uniform distribution (Figure 5) were identified on the basis of the elastic strain energy of a coplanar system of parallel cables and rubber [20].

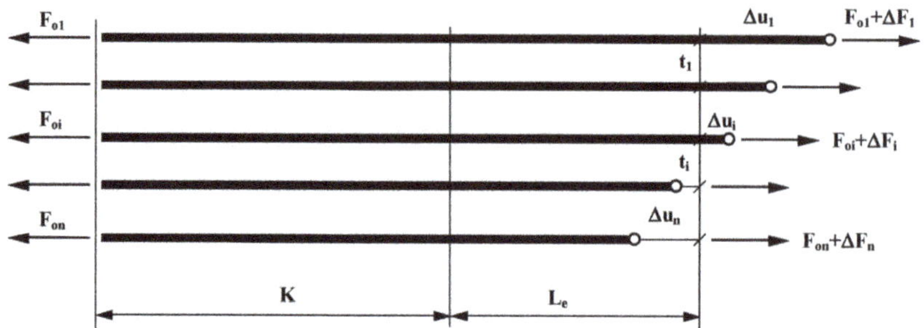

Figure 5. Schematic diagram of loads acting on cables in the transition section.

An increase in the elastic strain energy ΔE is accounted for by an increase in the elastic strain energy in the cable ΔE_l and by an increase in the elastic strain energy in the rubber ΔE_g.

$$\Delta E = \Delta E_l + \Delta E_g \tag{12}$$

Partial derivative of energy with respect to force provides the elongation as follows:

$$\frac{\partial \Delta E}{\partial \Delta F} = \Delta u \tag{13}$$

The analysis of steel-cord belt models [14] indicates that differential equations describing the variability of forces in the cords should be solved by combining hyperbolic functions, which is the effect of the character of differential equations. The development of these functions into Taylor series gives polynomials with strongly decreasing successive terms. Only the first two terms may be accepted with a satisfactory accuracy of several percent [3], and therefore, the equation describing the variability of forces in a single cord may be simplified to a quadratic function. This assumption is confirmed by the results of load tests performed for cords in the areas of the transition sections [13]. Due to the interaction between adjacent cables through the rubber layer (and in the case of textile belts, through the rubber and the weft threads), the force in any cable changes in accordance with the following relationship:

$$F(x) = F_o\left(1 + ax^2\right) \tag{14}$$

for $x = L_e + K = L$ as follows:

$$F_k = F_o\left(1 + aL^2\right) \tag{15}$$

where:

F_o—initial force, N;
F_k—final force, N.

Thus,

$$a = \frac{1}{L^2}\left(\frac{F_k}{F_o} - 1\right) \tag{16}$$

then we have the following:

$$F(x) = F_o + F_k\frac{x^2}{L^2} - F_o\frac{x^2}{L^2} \tag{17}$$

Increase in elastic strain energy in the cable is expressed with the following equation:

$$\Delta E_l = \frac{1}{2(EA)}\int_0^L F^2(x)dx = \frac{1}{2(EA)}F_o^2\int_0^L \left(1 + ax^2\right)^2 dx \tag{18}$$

$$\Delta E_l = \frac{1}{2(EA)} F_o^2 \left(L + \frac{2}{3}aL^3 + \frac{1}{5}a^2L^5 \right) \tag{19}$$

then, the partial derivative of elastic strain energy in the cable with respect to force is as follows:

$$\frac{\partial \Delta E_l}{\partial F_k} = \frac{\partial \Delta E_l}{\partial a} \cdot \frac{\partial a}{\partial F_k} \tag{20}$$

$$\frac{\partial \Delta E_l}{\partial F_k} = \frac{L}{(EA)} \left(\frac{3}{15}F_k + \frac{2}{15}F_o \right) \tag{21}$$

The next step consists of determining the elastic strain energy and derivative for the strip of rubber between the cables.

$$dE_g = \frac{1}{2}\Delta u(x)\tau h_t \cdot dx \tag{22}$$

where:

h_t—belt thickness, mm.

Tangential stresses τ are:

$$\tau = \gamma G^* = \frac{\Delta u(x)}{t}G^* \tag{23}$$

where:

G^*—artificial modulus of non-dilatational strain determined in laboratory tests, N/mm^2;
t—cable pitch, mm.

Then,

$$\Delta E_g = \frac{1}{2} \cdot \frac{h_t G^*}{t} \int_0^x \Delta u(x)^2 \, dx \tag{24}$$

Partial derivative of the energy in the ith rubber strip with respect to the force in the ith cable is as follows:

$$\frac{\partial \Delta E_{gi}}{\partial F_i} = \frac{h_t G^*}{t} \int_0^x \Delta u(x) \cdot \frac{\partial \cdot u(x)}{\partial F_i} \, dx \tag{25}$$

Partial derivative of energy with respect to force for the ith cable is as follows:

$$\frac{\partial \Delta E_i}{\partial F_i} = \frac{\partial \Delta E_{li}}{\partial F_i} + \frac{\partial \Delta E_{gi}}{\partial F_i} + \frac{\partial \Delta E_{gi-1}}{\partial F_i} \tag{26}$$

Relationships (21), (25), (13) provide a system of n equations of the following type:

$$A_{i-1} \cdot F_{i-1} + A_i \cdot F_i + A_{i+1} \cdot F_{i+1} + C_i = 0 \tag{27}$$

The matrix Gauss method allows a solution to the system of equations in the form of the vector of forces F_i in the cords.

4. Analysis of Loads on the Belt in the Region of the Transition Section in Troughed Conveyors

An analysis was conducted of loads acting on the belt in the transition section of a belt conveyor and of the influence of the belt type on the level of such loads involved for performing calculations for five conveyor belts having a strength of 2000 kN/m and various core designs. The calculations were performed for a steel-cord ST belt, an aramid core DP belt, a solid woven PWG belt, as well as for a multi-ply polyester–polyamide EP belt and a polyamide PP belt. Identical geometrical parameters of the transition section were used in each case.

The calculations were based on the physical and mechanical belt parameters identified in laboratory tests [21,22]: longitudinal modulus of elasticity E and artificial modulus of non-dilatational strain G^*. The calculations provided the distribution of loads on the cables/strips with respect to the belt width in the region of the pulley. Figure 6 shows force increments in the belt cables/strips in the cross section of the area where the belt enters the pulley, the increments being the effect of deformations in the transition section. The graph includes half of the belt thickness from the edge to the center.

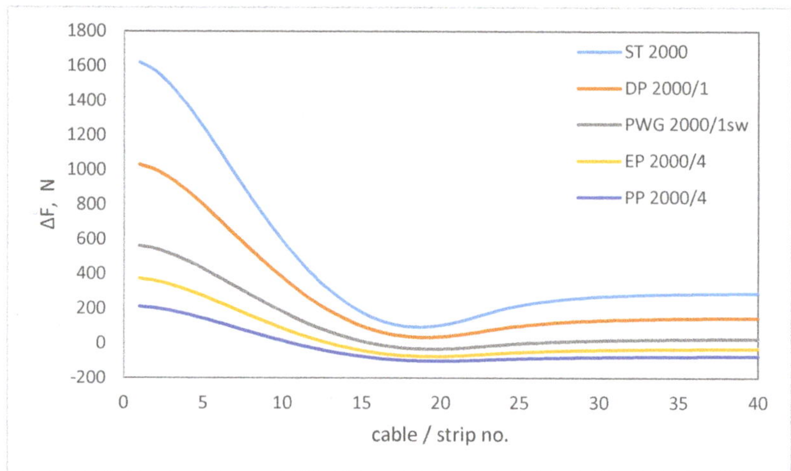

Figure 6. Load increment in the belt in the transition section of the belt conveyor (L_e = 2.6 m, B = 1.2 m, e = 80 mm, α = 45°).

In the analyzed cases, the absolute values of forces in individual belts are not important, as they result from the geometrical parameters of the transition section and its influence zone, as well as from the initial force in the belt, as assumed in the calculations. The behavior of different belts in the transition section can be reliably evaluated by observing the non-uniform character of loads acting on individual cables/strips in the belt. This non-uniform character of loads can be measured as a difference between the maximum and the minimum increment of force in the belt $\Delta F_{max} - \Delta F_{min}$. The load non-uniformity indicators thus obtained are shown in Table 2.

Table 2. Load non-uniformity indicators in the transition section.

Belt Type	$\Delta F_{max} - \Delta F_{min}$, in N
ST 2000	1525
DP 2000/1	995
PWG 2000/1 sw	592
EP 2000/4	446
PP 2000/4	312

The resulting load non-uniformity indicators are significantly different for the analyzed belts. As compared to the multiply polyamide belt, the load non-uniformity indicators for the polyester–polyamide belt are 50% higher, and for the solid woven, aramid core and steel-cord belts, it is two-fold, three-fold and five-fold higher, respectively. Data presented in manuals provided by the most prominent conveyor belt manufacturers and in standards define relationships for adjusting the parameters of the transition section and the belt, which are based on simplified equations allowing for the geometry and the longitudinal modulus of elasticity of the belt, but not for the interactions between the adjacent cables/strips. The approach is different solely in the case of designing the transition section for a steel-cord belt and for the remaining group of textile belts. Importantly, however, when replacing a worn belt, each decision to change the belt type should be preceded by an analysis aiding the selection of the geometrical parameters of the transition section in the belt conveyor. This is of particular importance in the location where the greatest longitudinal forces are observed in the belt, i.e., typically in the drive pulley. In the case when the operator decides to install a belt having different elastic properties but without modifying the geometrical parameters of the transition section, the limit forces in the cables/strips can be exceeded, and a cascading break may develop in the belt.

5. Conclusions

Breaks in a belt loop operated on a conveyor develop in the splice, and very rarely in the "continuous" part of the belt. The aim of this article is not to modify any good practices developed in the context of designing and operating belt transportation, but rather to indicate the so-called "sensitive points."

Importantly, belt tensile strength tests indicate that an increase in the specimen width entails a decrease in the belt strength (Figure 1). If the test method is assumed not to have an influence on this phenomenon, it may be attributed to the specific construction of the core in conveyor belts. In structures, such as the belt—with multiple plies and fibers—each thread may have a various preliminary load, due to, for example, preliminary tension, weave system or thermal shrinkage, which occurs during the belt production process. In such a case, the breaking process is gradual: the threads with the highest preload break first and the remaining threads, later. As a consequence, the strength will be lower than the multiple of the strength of a single thread. As both the safety factor used in the process of selecting belt strengths and the belt strength level observed in the tests are relatively high, the above phenomenon seems appropriate. This aspect is nevertheless important when using belts of great widths, which pass curvature geometries along the route and the conveyor transition sections, especially in the zone of the highest specific forces in the belt. Due to constraints of a technical nature, the tests were performed on multi-ply low-strength belts. Recommendations for further research include performing similar tests also for belts of different core design and verifying the test results with the use of computer simulation methods.

The transition section of the conveyor is a region in which the forces in the belt cross-section vary significantly. Such non-uniformity of belt loads was confirmed in the analysis performed with the use of an original theoretical model of the belt in the transition section of the troughed conveyor, which allows for the elastic properties of the belt and for the interaction with adjacent cables/strips. A comparison of the obtained results with the FEM test results obtained by other authors [17] indicates significant similarity of the stress patterns across the belt width. The lowest load non-uniformity was observed in the belt with multi-ply polyamide core, and higher values—in belts with polyester–polyamide core, with solid woven core, with aramid core and with steel-cord core, respectively. When replacing a worn belt, each decision to change the belt type should be preceded by a new analysis of the geometrical parameters of the transition section in the belt conveyor. In the case when the operator decides to install a belt having different elastic properties but without modifying the geometrical parameters of the transition section, the belt load non-uniformity may be increased by as much as several hundred percent (Table 2, Figure 6). This fact is of special importance in the case of the transition section before the drive pulley, where typically, the highest force levels in the belt are observed.

Author Contributions: Conceptualization, D.W. and M.H.; methodology, D.W.; software, D.W.; validation, M.H.; formal analysis, M.H. and D.W.; investigation, D.W.; resources, M.H. and D.W.; data curation, D.W.; writing—original draft preparation, D.W.; writing—review and editing, M.H. and D.W.; visualization, D.W.; supervision, M.H.; project administration, D.W. All authors have read and agreed to the published version of the manuscript.

Funding: This research work was co-founded with the research subsidy of the Polish Ministry of Science and Higher Education granted for 2021.

Institutional Review Board Statement: Not applicable.

Informed Consent Statement: Not applicable.

Data Availability Statement: The data presented in this study are available on request from the corresponding author.

Conflicts of Interest: The authors declare no conflict of interest.

References

1. Hardygóra, M.; Wachowicz, J.; Czaplicka-Kolarz, K.; Markusik, S. *Taśmy Przenośnikowe. (Conveyor Belts)*; WNT: Warszawa, Poland, 1999; ISBN 83-204-2402-X. (In Polish)
2. Lodewijks, G. *Dynamics of Belt Systems*; TU Delft: Delft, The Netherlands, 1996; ISBN 90-370-0145-9.
3. Gładysiewicz, L. *Przenośniki Taśmowe. Teoria i Obliczenia. (Belt Conveyors. Theory and Calculations)*; Wroclaw University of Technology Publishing House: Wroclaw, Poland, 2003; ISBN 83-7085-737-X. (In Polish)
4. Bajda, M. Laboratory tests of conveyor belt parameters affecting its lifetime. In Proceedings of the 17th International Multidisciplinary Scientific GeoConference SGEM, Sofia, Bulgaria, 29 June–5 July 2017; Book 13, pp. 495–502. [CrossRef]
5. Bajda, M.; Błażej, R.; Jurdziak, L. A new tool in belts resistance to puncture research. *Min. Sci.* **2016**, *2*, 173–182. [CrossRef]
6. Andrejiova, M.; Grincova, A.; Marasova, D. Analysis of tensile properties of worn fabric conveyor belts with renovated cover and with the different carcass type. *Maint. Reliab.* **2020**, *22*, 472–481. [CrossRef]
7. Andrejiova, M.; Grincova, A.; Marasova, D. Failure analysis of the rubber-textile conveyor belts using classification models. *Eng. Fail. Anal.* **2019**, *101*, 407–417. [CrossRef]
8. Köken, E.; Lawal, A.I.; Onifade, M.; Özarslan, A. A comparative study on power calculation methods for conveyor belts in mining industry. In *International Journal of Mining, Reclamation and Environment*; Taylor & Francis Group: Abingdon, UK, 2021. [CrossRef]
9. CEMA5 Conveyor Equipment Manufacturers Association. *Belt Conveyors for Bulk Materials*, 5th ed.; Conveyor Equipment: Naples, FL, USA, 2002; ISBN 1-891171-18-6.
10. Dunlop, F. *Conveyor Handbook*; Conveyor Belting Australia: Caboolture, Australia, 2009; p. 103.
11. German Standard DIN 22101. *Continuous Conveyors—Belt Conveyors for Loose Bulk Materials—Basis for Calculation and Dimensioning*; Deutsches Institut fur Normung E.V. (DIN): Berlin, Germany, 2011.
12. Oehmen, K.H. Berechnung der Dehnungsverteilung in Fördergurten infolge Muldungsübergang, Gurtwendung und Seilunterbrechung. *Braunkohle* **1979**, *31*, 394–402.
13. Hager, M.; Tappeiner, S. Additional Strain in Conveyor Belts Caused by Curves and Transition Geometry. *Bulk Solids Handl.* **1993**, *13*, 695–703.
14. Schmandra, A. Allgemeines Modell für die Berechnung von Spannungsverläufen in Stahlseilfördergurten. *Hebezeuge Fordermittel Berl. 31* **1991**.
15. Harrison, A. Modelling belt tension around a drive drum. *Bulk Solids Handl.* **1998**, *18*, 75–80.
16. Fedorko, G.; Ivančo, V. Analysis of force ratios in conveyor belt of classic belt conveyor. *Procedia Eng.* **2012**, *48*, 123–128. [CrossRef]
17. Fedorko, G.; Beluško, M.; Hegedűš, M. FEA utilization for study of the conveyor belts properties in the context of internal logistics systems. In Proceedings of the Carpathian Logistics Congress, CLC, Jesenik, Czech Republic, 4–6 November 2015; Tanger Ltd.: Ostrava, Czech Republic, 2015; pp. 293–299, ISBN 978-80-87294-61-1.
18. Mikušová, N.; Millo, S. Modelling of Conveyor Belt Passage by Driving Drum Using Finite Element Methods. *Adv. Sci. Technol. Res. J.* **2017**, *11*, 239–246. [CrossRef]
19. Blazej, R.; Jurdziak, L.; Kawalec, W. Operational Safety of Steel-Cord Conveyor Belts under Non-stationary Loadings. In *Advances in Condition Monitoring of Machinery in Non-Stationary Operations*; CMMNO Applied Condition Monitoring; Springer: Cham, Switzerland, 2014; Volume 4, pp. 473–481. [CrossRef]
20. Woźniak, D. Wpływ Konstrukcji Taśmy Oraz Geometrii Odcinka Przejściowego Na Rozkład Obciążeń W Taśmie Z Linkami Stalowymi. (Influence of Belt Design and Transition Section Geometry on Load Distribution in Steel Cord Belt). Ph.D. Dissertation, Wroclaw University of Science and Technology, Wrocław, Poland, 1998. (In Polish)
21. Gładysiewicz, L.; Woźniak, D. Theoretical model of steel cable belt in the transitory zone of a pipe conveyor. *Sci. Pap. Inst. Min. WUT No. 83* **1998**, *22*, 64–74.
22. Woźniak, D.; Sawicki, W. Doświadczenia z badań laboratoryjnych modułu sprężystości taśm przenośnikowych przy cyklicznym rozciąganiu. (Experiences from laboratory tests of conveyor belt elasticity modulus upon cyclical loading). *Transp. Przemysłowy Masz. Rob.* **2008**, *2*, 6–10. (In Polish)

Article

Models of Transverse Vibration in Conveyor Belt—Investigation and Analysis

Piotr Bortnowski, Lech Gładysiewicz, Robert Król and Maksymilian Ozdoba *

Department of Mining Wroclaw University of Science and Technology, Na Grobli 15, 50-421 Wroclaw, Poland; piotr.bortnowski@pwr.edu.pl (P.B.); lech.gladysiewicz@pwr.edu.pl (L.G.); robert.krol@pwr.edu.pl (R.K.)
* Correspondence: maksymilian.ozdoba@pwr.edu.pl

Abstract: The transverse vibration frequency of conveyor belts is an important parameter describing the dynamic characteristics of a belt conveyor. This parameter is most often identified from theoretical relationships, which are derived on the basis of an assumption that the belt is a stationary elastic string. Belt vibrations have a number of analogies to other tension member systems, such as, for example, power transmission belts. Some research findings suggest that in the case of a limited length of the belt section, a more accurate description of its vibration can be obtained with a beam model rather than with a string model. Experimental research has so far mostly revolved around measurements of stationary belts. This article presents the results of vibration measurements performed for a moving belt and obtained for various operating parameters of the conveyor, as well as for several configurations of the distance between idler supports. The analysis was conducted on a moving steel-cord belt. Belts of this type are commonly used in the majority of mines and industrial plants. The measurement results were compared with the model of a string and with the model of a beam in tension. Both of the theoretical models allowed for the belt speed, whose influence was demonstrated in both theoretical calculations and experimental tests to be negligible. On the other hand, the tensile force in the belt was observed to have a significant impact on the vibration frequency. Depending on the idler spacing, the measurement results are approximate to those of the beam model or of the string model. For spacing smaller than 1.6 m, the belt shows properties approximate to an elastic beam, while for spacing greater than 1.6 m, the belt behaviour can be better represented through a string model. A beam model is, therefore, more applicable in analyses of vibrations in the upper strand of the belt, while a string model is more useful in analyses of vibrations in the lower strand.

Keywords: conveyor belt; transverse vibrations; laboratory and in-service tests of conveyor components; noise emissions

Citation: Bortnowski, P.; Gładysiewicz, L.; Król, R.; Ozdoba, M. Models of Transverse Vibration in Conveyor Belt—Investigation and Analysis. *Energies* **2021**, *14*, 4153. https://doi.org/10.3390/en14144153

Academic Editor: Wiseman Yair

Received: 31 May 2021
Accepted: 6 July 2021
Published: 9 July 2021

Publisher's Note: MDPI stays neutral with regard to jurisdictional claims in published maps and institutional affiliations.

1. Introduction

The analysis of transverse vibrations of the conveyor belt is important, due to their destructive effect on the durability of structural elements and due to the areas of resonant work [1–3]. Resonances can result in excessive wear of the belt and idlers, but also increased noise emission. The regulations define the noise emission requirements of machinery and equipment approved for industrial use [4–6]. In recent years, these requirements have become increasingly restrictive. In particular, open-cast mines where conveyor belts are the main transport solution must enact a series of solutions to reduce noise emissions to the environment [7–10]. It is necessary to look for new solutions, and one of them may be the analysis of transverse vibrations of the conveyor belt in terms of their impact on the noise emission of the conveyor. Research on the transverse vibrations of elastic belts, including conveyor belts, has been conducted for a long time [1,11–14]. In most publications, the main research objective is to find the vibration frequency of the belt based on analytical methods [1,15–17]. Laboratory tests are usually limited to measurements under static conditions of a stationary belt tensioned with constant force [18]. The frequency of the tape vibrations depends on parameters such as stiffness, tension force, linear movement speed, mass, section geometry, and number and spacing of support sets [19]. Some of these parameters can be determined by measuring on a conveyor or by examining belt samples in the laboratory [20,21]. With a good theoretical model, these parameters can accurately predict the frequency of belt vibrations in the initial stages of conveyor design. The motivation for research and consideration of the form of current models was the fact that they describe the frequency of vibrations with a high approximation, using simplified physical models as a base, and there are no studies verifying their usefulness in dynamic conditions [19]. The basis for the calculation in the case of a conveyor belt is the string model, which, in practice, aims to verify that the forces in the upper or lower empty conveyor do not fit the resonance force in terms of changes in the forces in the upper or lower empty conveyor [22]. Some studies suggest that the tape shows the characteristics of the string, but only in the case of large spacings of support sets [23]. For such conditions, it is necessary to use a different model—the model of the beam freely supported. This model completely ignores the dynamics of the conveyor belt [19]. The belt should be analysed as a moving object, as this corresponds to the working conditions. The model of the tensile beam is further proposed as a better approximation of the behaviour of the movable guide, with increased stiffness. Additional parameters have also been introduced into the models: the effective force in the tape, a linear velocity, and a change in the tensile force of the belt. The theoretical results obtained were verified using a non-contact measuring method of the mobile belt using a vibration and noise meter. A detailed analysis of the model errors and the impact of the individual conveyor parameters on the result obtained is provided.

2. Materials and Methods

2.1. Theory and Tests of Conveyor Belt Transverse Vibrations

The belt is a tension member, and therefore, a number of analogies can be found between its behaviour and transverse vibrations of power transmission belts [22–26]. Thus, the results of investigations into belt drive vibrations may be useful in analyses of conveyor belt vibrations. Such investigations of power transmission belts are based on models that represent frictional engagement between the tension member and the belt pulley [1,15,16,27–29]. In stationary conditions, when the tension member is considered as an elastic string, the expression describing the vibration frequency is as follows [19]:

$$f_s = \left(\frac{n \cdot \pi}{2 \cdot L}\right) \cdot \sqrt{\frac{T}{m}} \ (\text{Hz}), \tag{1}$$

where: n—natural number (base vibration frequency is calculated for $n = 1$), L—string length (m), T—tension force (N), and m—linear mass of the tension member (kg/m).

Unlike a string, the conveyor belt has a defined width B and extends a length l, which corresponds to the spacing between the idler sets. A commonly used parameter assigned to a particular belt is its surface weight m_t. Equation (1) can be thus represented as follows:

$$f_s = \left(\frac{n \cdot \pi}{2 \cdot l}\right) \cdot \sqrt{\frac{T}{B \cdot m_t}} \ (\text{Hz}), \tag{2}$$

where: B—belt width (m), and m_t—unit surface weight of belt (kg/m^2).

With allowance for the movement of the belt with a speed v on a curve, which is a sag line between two idler supports, an equivalent force in the belt can be introduced [2,30]:

$$T_0 = T - B \cdot m_t \cdot v^2 \ (\text{N}) \tag{3}$$

and thus, the vibration frequency is:

$$f_s = \left(\frac{n \cdot \pi}{2 \cdot l}\right) \cdot \sqrt{\frac{T - B \cdot m_t \cdot v^2}{B \cdot m_t}} \ (\text{Hz}), \tag{4}$$

where: v—belt speed (m/s); l—length between the idler sets (m).

The theoretical string model does not include the elastic and damping parameters, and this fact may be of importance in the case of a conveyor belt. Model [19] allows for the equivalent stiffness of the system by introducing the θ coefficient and includes a modified form of Equation (1) representing base vibration frequency (for n = 1):

$$f_s = \frac{\pi}{2 \cdot L} \cdot \sqrt{\frac{T}{m}} \cdot \left[1 - \theta \cdot \frac{v^2}{c^2}\right] \text{ (Hz)}, \tag{5}$$

where: c—speed of elastic wave in the belt (m/s), and θ—dimensionless stiffness coefficient of the system.

In Equation (5), the θ coefficient assumes a value from a range from 0 to 1, and elastic wave velocities in belts are more than tenfold higher than the belt velocities [18,31–33]. Therefore, an additional element of Equation (5), which expands relationship (1), tends to one and, in practice, is equal in form to the relationship for a stationary string. The application of Equation (5) requires finding the θ system stiffness coefficient, which is not easily identified for a belt installed on a conveyor.

The string model does not allow for the influence of material properties on the vibration frequency in the belt. The transverse vibration frequency is correlated with the propagation speed of the elastic wave in the conveyor belt, and this depends on the longitudinal modulus of elasticity and on the rheological parameters. An assumption can be made, therefore, that belt transverse vibrations depend on the above material parameters [2,33]. In some cases, models of vibration frequency in elastic tension members (including conveyor belts) should include flexural rigidity [34,35]. Belt elastic properties and flexural rigidity can be allowed for by using a model of a stationary simply supported beam having a given vibration frequency [19,36]:

$$f_b = \frac{n \cdot \pi}{2 \cdot l^2} \cdot \sqrt{\frac{EJ}{B \cdot m_t}} \text{ (Hz)}, \tag{6}$$

where: EJ—flexural rigidity (Nm^2).

As it is a simply supported beam, the tensile force is not allowed for in the equation. With allowance for the tensile force, the beam will have its base vibration frequency as follows:

$$f_b = \frac{n \cdot \pi}{2 \cdot l^2} \cdot \sqrt{\frac{EJ}{B \cdot m_t}} \cdot \sqrt{\left(1 + \frac{T \cdot l^2}{\pi^2 \cdot EJ}\right)} \text{ (Hz)}. \tag{7}$$

In a manner analogous to the string, equivalent force in the belt can be introduced in Equation (6), and thus, the resultant relationship allows for belt speed:

$$f_b = \frac{n \cdot \pi}{2 \cdot l^2} \cdot \sqrt{\frac{EJ}{B \cdot m_t}} \cdot \sqrt{\left[1 + \frac{(T - B \cdot m_t \cdot v^2) \cdot l^2}{\pi^2 \cdot EJ}\right]} \text{ (Hz)}, \tag{8}$$

The necessity to investigate theoretical models other than the string model is suggested in [37], in which differences from the string model were observed when the support spacing was limited. The above research employed a contactless device for measuring base eigenfrequency of vibrations in a power transmission belt. The main objective of the tests was to find a relationship between the vibration frequency and the belt tensioning. The study demonstrated such a possibility, albeit not for short-span belts, in which the string model did not allow precise results.

Except for the analytical models, analyses of vibrations in tension members, e.g., in power transmission belts, are based on numerical methods that employ the Finite Element Method (FEM) [38–40]. Numerical methods offer highly accurate results, but only in the case when both the boundary conditions of the systems and the material parameters are well defined. However, their precise identification is not always possible; therefore, laboratory tests still remain the basic source of information on vibrations in tension members. The modelling of transverse vibrations in conveyor belts is additionally hindered by the non-uniform cross-section of the belt and by the wide variability range of tensile forces when the belt is operated on the conveyor. Moreover, in the case of the upper strand, an allowance should be made for the variability of transverse loads due to the belt own mass, the mass of the transported material, and the belt movement.

Research described in [18] presents identifications of eigenfrequencies in a stationary conveyor belt with the use of a step response method. A pretensioned belt was installed in a device for testing dynamic characteristics and for exciting vibrations by cyclical impacts of known frequency. Sensors located in selected points on the belt recorded the response signal, and this was used to determine the stress wave speed and the belt vibration frequency. The idea behind the measurement rig was to allow the use of displacement or acceleration sensors. The tests were performed for two types of belts, 3.2 m in length and 0.25 m in width, and for belt tension force from 6 kN to 20 kN. Within the investigated range of tensile forces, and for a steel-cord belt having a nominal strength of 1000 kN/m, the measured mean eigenfrequency was 14.45 Hz, and for a textile belt having a strength of 800 kN/m, the mean eigenfrequency was 11.75 Hz. In the case of both belts, the frequency increase was observed to be non-linear with respect to the increase in the belt tension force.

The literature offers a limited number of publications on measuring transverse vibration in a moving conveyor belt. A belt in motion is a significant complication in measurements. One of the proposed methods is based on contactless electrostatic sensors, which react to both longitudinal and transverse belt movement [41]. This method has been developed mainly with a view of continuous measurements of vibrations for diagnostic purposes and for predicting the behaviour of belt drives in industrial conditions. The tests indicated that the signal from the sensor is dominated by the transverse vibrations of the belt. Another measurement method consisted in using images from a high-speed industrial camera to identify transverse vibration in a power transmission chain [42]. The lens of the camera followed markers positioned on the moving chain, and the image was continuously processed with the use of computer software.

2.2. Equipment and Methodology

The object of the tests was a conveyor belt with the core comprising steel cords and with a nominal tensile strength of 800 kN/m. A steel belt is commonly used in most opencast mines, where noise is particularly annoying [22]. The nominal belt strength was limited by the drive power of the test rig and the dimensions of the drums. The belt parameters important for calculating theoretical vibration models are presented in the table below (Table 1).

Table 1. Parameters of the tested belt ST800.

Width B (m)	0.435
Loop length L_t (m)	11.78
Tensile strength K_n (kN/m)	800
Core type	steel
Thickness h_t (m)	0.018
Flexural rigidity EJ (Nm^2)	4.49
Mass mt (kg/m^2)	23.3

The flexural rigidity EJ was calculated in a simple manner by measuring the sag of the free end of a belt sample restrained on one end. The measured sag was due to only the own mass of the belt.

The laboratory test rig consists of two pulleys—the drive pulley and the return pulley, spaced 5 m from each other. The upper strand can be supported by up to three idler supports (the idler dimensions being ϕ130 × 500) with adjustable spacing between the idlers. The return pulley is tensioned with the use of two hydraulic actuators. Hydraulic regulation allows for stable tensile force in steady motion. This fact is advantageous because the frequency of transverse vibrations depends on the tensile force in the belt. The measurements of the belt were performed with the use of two sensors installed on the two sides of the tensioning pulley KM1603, with a measuring range equal to 50 kN, and with a maximum relative linearity error at 0.03%. The sensors provided measurements with a sensitivity of 2 mV/V (for a nominal power supply of 5 V). Belt speed was measured with incremental encoders with an accuracy of up to 0.02 m/s [43]. Figure 1 shows the test rig with indicated locations of idler supports.

Figure 1. The test rig with indicated locations of idler supports in the upper strand.

The tests were performed with the use of the directional sound and vibration meter SVAN979 by Svantek ltd (Svantek, Warsaw, Poland), which was positioned in the vicinity of the upper strand, in the centre, between the idler supports. The head of the meter allows measurements from 0.5 Hz. Belt vibrations cause cyclical changes of acoustic pressure, which are the input signals in the meter. A wide measurement range of the meter (from 0.5 Hz to 20 kHz) and the sensitivity of the head (4 mV/Pa) allow for the recording of local changes of acoustic pressure within the area of the vibrating belt. The frequency of these changes corresponds to the belt vibration frequency. The sensor located in the centre (central point between the supports) allows the highest accuracy of the measurement because records are made of maximum vibration amplitudes. The directional operation of the measuring head is of particular importance, as the recorded changes are only those due to belt vertical movement. The signal from the meter was sampled at a frequency of 10 kHz, with the use of the NI USB-4432 data acquisition system. The signal was further processed in LabView. Figure 2 is the diagram of the measurement system.

The final stage in the processing of the results was to use signal analysis tools in order to identify the base frequency of transverse vibrations in the belt (Figure 3). Theoretical equations and the known parameters of the belt served as a basis to calculate the expected frequency ranges in which the belt vibration frequency is located. Preliminary identification of such ranges is advantageous in selecting proper frequencies. The test rig was operated in an acoustically isolated room, which allowed clear spectra of the recorded signal. Based on the distance between the meter and the belt, the spectrum can be assumed to be dominated by the frequencies related to the belt transverse vibrations. The theoretically identified areas were searched for frequencies of dominant amplitudes (Figure 3b). The input signal (Figure 3a) was subsequently filtered with the use of a band-pass filter, whose parameters were defined on the basis of a prior spectral analysis. The result was a time signal, which could be directly related to the amplitude of the belt transverse vibrations (Figure 3c). The analysis of the spectrum of the filtered signal allowed for the isolation of those areas in which the dominant frequencies are related to belt transverse vibrations, along with the neighbourhoods in which the noise is low depending on the parameters of the applied filter (Figure 3d).

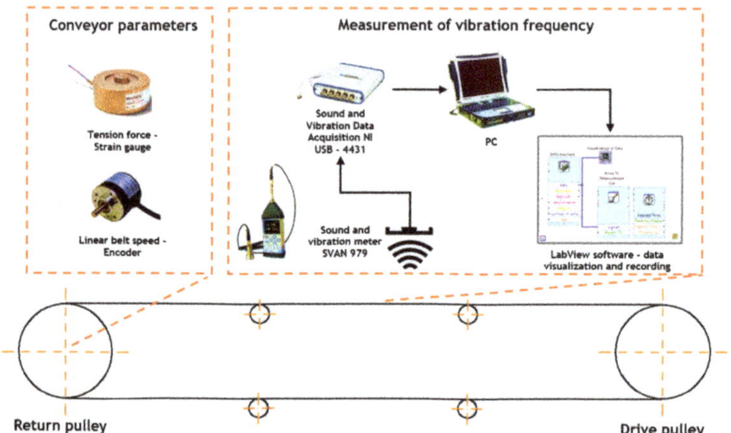

Figure 2. Diagram of the measurement systems—the system for measuring the basic operating parameters of the conveyor (on the left), and the system for measuring transverse vibration frequency in the belt (on the right).

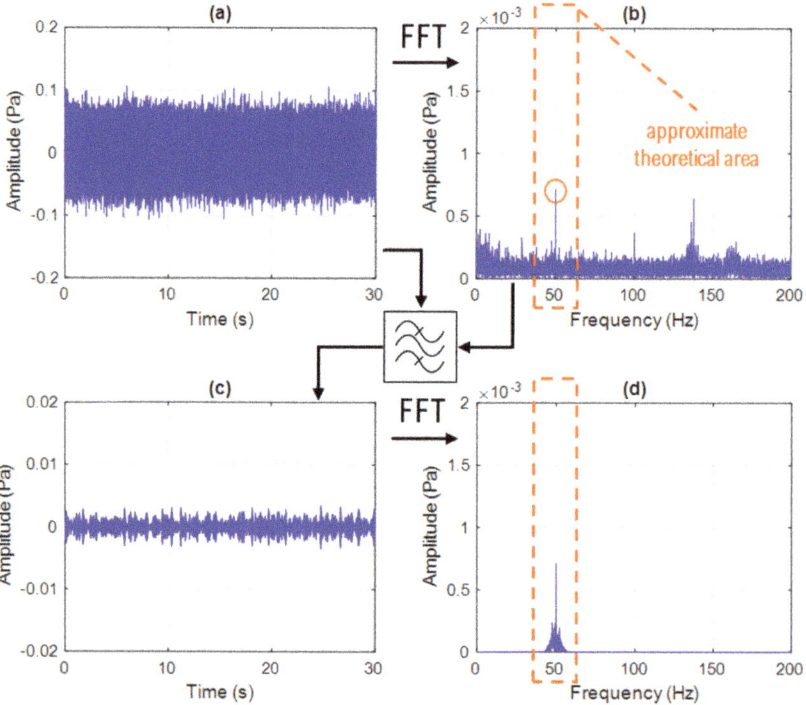

Figure 3. Diagram representing the isolation of belt vibration frequency from the recorded signal: (**a**) signal recorded from the measuring head above the belt, (**b**) signal spectrum above the belt, (**c**) filtered frequencies located within a close range to the frequencies of belt transverse vibrations, and (**d**) spectrum of the filtered signal.

Despite the lack of direct contact between the measuring head and the moving belt, the transverse vibration frequencies are visible in the obtained frequency characteristics. In the case of greater distances (Figure 4a), relatively high vibration amplitudes are observed, which translate into significant changes of acoustic pressure, as recorded by the sensor. In the case of small distances, the obtained spectra are less clear, and the concentration of noise in the very low frequency range hinders the identification of the proper frequency of belt transverse vibrations (Figure 4b). In this case, the interpretation of the results requires a close analysis of the spectrum.

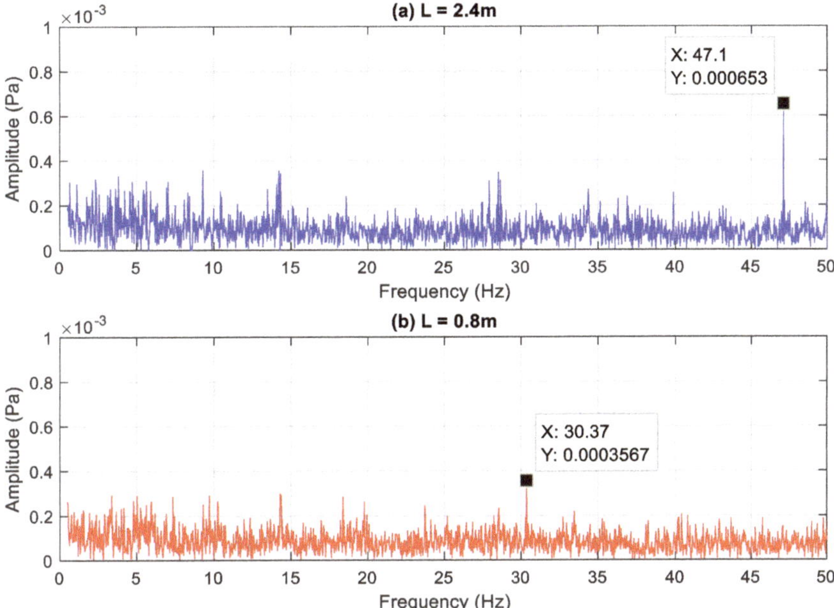

Figure 4. Spectrum of the acoustic signal from the meter for a belt linear speed of 2.4 m/s and for a belt tensile force of 40 kN: (**a**) idler spacing at 2.4 m; (**b**) idler spacing at 0.8 m.

3. Results

The test plan involved changes of the three parameters of influence: belt tensile force, belt speed, and idler support spacing. The travelling of the hydraulic actuators, which move the return pulley on the test rig, allows for adjustments of forces in the belt within a range from 20 kN to 60 kN. Five measurement series were performed at 10 kN intervals. The maximum belt speed is limited by the parameters of the drive system and is approx. 4 m/s. The tests were performed for 5 belt speeds, in the range of 20% to 100% of the maximum speed allowed for by the drive system. In the case of the measurement series in which the idler support spacing was changed from 0.8 m to 2.4 m, the modifications were inspired by typical spacing values used in practice [44]. These 3 parameters are the basic information about the working condition of the conveyor. Additionally, their significance is confirmed by theoretical equations. Presumably, the spacing of the idler supports can have a significant effect on the belt stiffness, which is important for analytical calculations.

The influence of the belt speed was observed to be negligible. This fact is demonstrated in the graph of Figure 5. In this particular case, for L = 1.2 m, the measurement results are similar to the beam model and are significantly different than the string model, albeit with areas in which reverse proportions are found. Both the theoretical models and the measurement results indicate a minimal, practically negligible decrease of the belt vibration frequency in relation to the belt speed. For this reason, the influence of belt speed is not considered in further analyses.

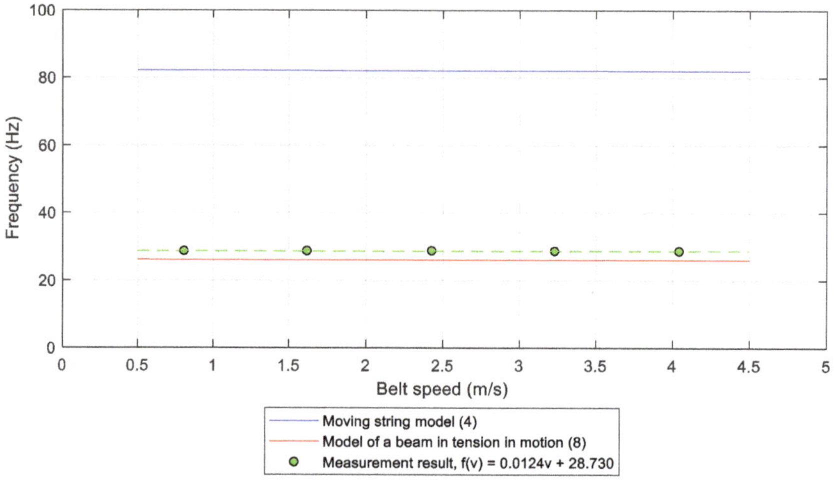

Figure 5. The frequency of transverse vibrations as a function of belt speed, for support spacing L = 1.2 m compared with the results predicted by the theoretical models of a string and of a beam (for a tensile strength of 40 kN).

Figure 6 shows the frequencies of belt transverse vibrations as a function of belt tensile force for two selected idler spacing values (L = 1.2 m—Figure 6a and L = 2 m—Figure 6b). The two diagrams show both the theoretical relationships for the two analysed models and the averaged measurement points. The theoretical relationships are power functions in the form of y = ax^(b), where the exponent is 0.5, and therefore, the set of the measurement points was approximated with curves having an identical form. As a result, high correlation coefficients of empirical functions $R2$ were obtained. An analogical procedure was performed for the remaining measurements. The approximations of the measurement results were used to identify the differences between these functions and the theoretical relationships. The differences between the theoretical models and the measurement results are shown in Table 2. The table presents Mean Absolute Error (MAE—the error of the model in relation to the measurement result) for all of the investigated idler spacing values.

Both models differ from the measurement results less within the range of small forces (20 kN) than within the range of great forces (60 kN).

Table 2. Differences between the theoretical models and the mean measurement results for the belt speed of 2.4 m/s.

Idler Spacing L [m]	Method		Tension Force T (kN)									
			20		30		40		50		60	
	Measurement (Hz)		28.6		32.4		47.1		50.1		58.8	
2.4	MAE (Hz)/(%)	String (4)	0.4	1.5%	3.1	9.7%	6.0	12.8%	4.2	8.3%	8.5	14.4%
		Beam (8)	19.4	67.7%	21.1	65.1%	34.0	72.2%	35.5	70.8%	42.8	72.8%
	Measurement (Hz)		33.1		35.9		44.4		47.0		51.0	
2.0	MAE (Hz)/(%)	String (4)	1.8	5.3%	6.8	19.0%	4.9	11.0%	8.1	17.3%	9.4	18.4%
		Beam (8)	22.0	66.5%	22.3	62.1%	28.7	64.7%	29.4	62.7%	31.8	62.3%

Table 2. *Cont.*

Idler Spacing L [m]	Method		Tension Force T (kN)									
			20		**30**		**40**		**50**		**60**	
	Measurement (Hz)		**20.0**		**28.8**		**44.6**		**50.0**		**59.6**	
1.6	MAE (Hz)/(%)	String (4)	23.6	118.1%	24.6	85.3%	17.1	38.3%	18.9	37.7%	15.9	26.6%
		Beam (8)	6.1	30.6%	11.8	41.0%	24.9	56.0%	28.1	56.1%	35.6	59.7%
	Measurement (Hz)		**14.4**		**19.9**		**28.8**		**36.8**		**38.7**	
1.2	MAE (Hz)/(%)	String (4)	43.7	303.2%	51.2	257.0%	53.3	185.0%	55.1	149.7%	62.0	160.3%
		Beam (8)	4.1	28.4%	2.7	13.7%	2.7	9.2%	7.5	20.5%	6.6	17.1%
	Measurement (Hz)		**18.4**		**22.1**		**30.4**		**36.8**		**39.9**	
0.8	MAE (Hz)/(%)	String (4)	68.7	372.6%	84.6	383.5%	92.9	305.9%	101.0	274.5%	111.1	278.4%
		Beam (8)	9.3	50.7%	11.9	54.1%	8.9	29.3%	7.1	19.3%	8.2	20.5%

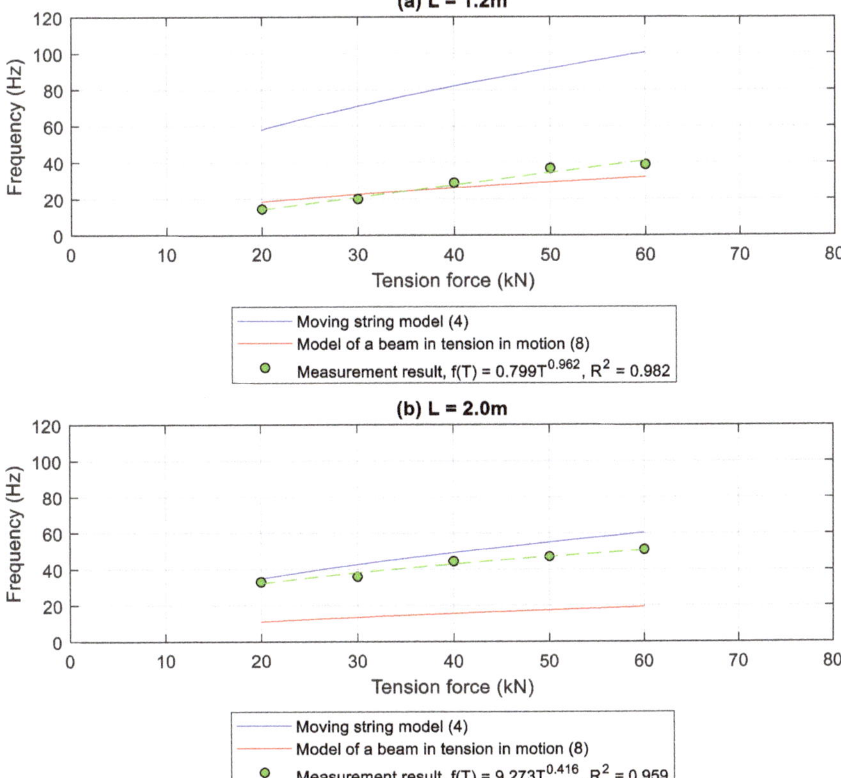

Figure 6. Comparison of the measured frequencies of belt transverse vibrations with the theoretical models as a function of belt tensile force for the belt speed of 2.4 m/s: (**a**) idler spacing L = 1.2 m; (**b**) idler spacing L = 2.0 m.

4. Discussion

The results shown in the diagrams of Figure 6 and in Table 2 do not provide a definite answer to the question of which of the two analysed models better represents belt transverse vibrations. Areas convergent to both the string model and the beam model are identified depending on the distance between the idler supports. The 3D plot in Figure 7 shows a surface resulting from the approximation of the measurement results of belt transverse vibration frequencies as a function of tensile force and idler spacing. The surface was obtained by quadratic interpolation of the points identified in the measurements. In accordance with the theoretical relationships, the surface for the two models should be monotonically increasing together with an increase in force and monotonically decreasing together with an increase in support spacing. However, some areas clearly show differences from the theoretical assumptions, manifested in local minima and maxima on the surface. This fact may be accounted for by the diagram shown in Figure 8, which is plotted for the analysis of the influence of the idler spacing.

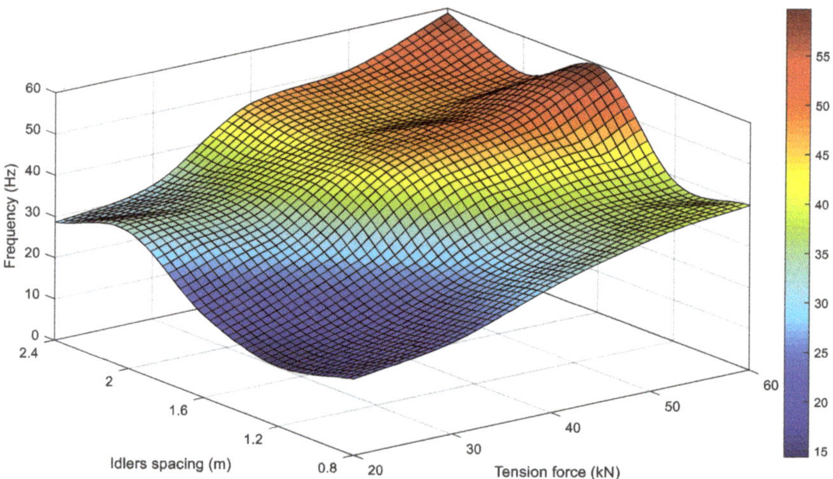

Figure 7. Frequency of belt transverse vibrations as a function of two measured parameters: belt tensile force and spacing between the idler supports.

This analysis was based on the mean force of 40 kN obtained in the measurements and it represents the influence of idler spacing on the frequency of belt transverse vibrations. Five averaged measurement points for this level of force together with the confidence intervals were compared with the theoretical relationships (Figure 8). The area indicated with number 3 in Figure 8 shows similarities between the measurement results and the string model for high idler spacing values, while the area indicated with number 1 shows similarities with the beam model for low idler spacing values. Point 2 in Figure 8 is located on the boundary between the two areas. For the entire range of forces in the belt, the set of points located on the boundary of the areas similar to the string model and to the beam model form local minima on the surface of the 3D plot (Figure 7).

As a result, a certain idler spacing range is observed, in which both the string model and the beam model have a significant error. MAE in relation to the two models can be, therefore, used to analytically determine the limit idler spacing for the application of a particular model (Figure 9). This limit spacing is indicated by the intersection point of two straight lines obtained by approximating the MAE.

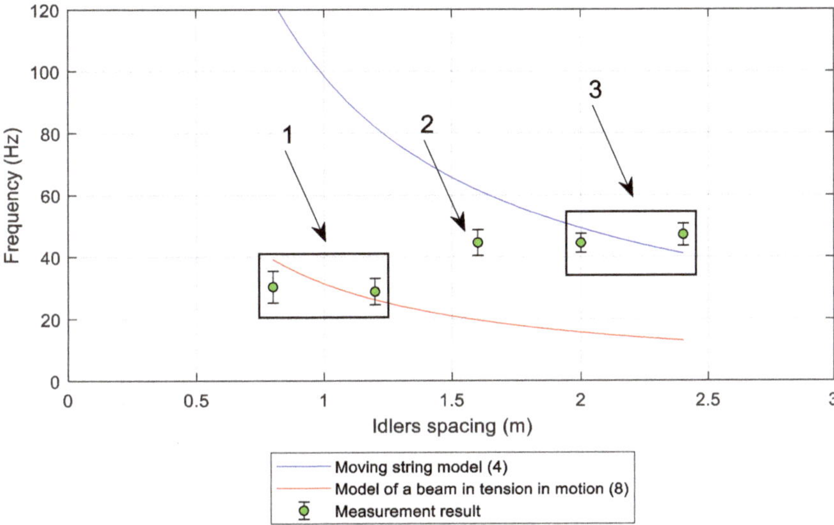

Figure 8. The measured frequency of transverse vibrations in the ST800 belt as a function of idler support spacing for a constant belt linear speed (2.4 m/s) and for a constant tensile force (40 kN), compared with the values predicted for the moving string model and for the model of a moving beam in tension: 1—area of optimal prediction from the beam model, 2—intermediate point with a significant error from both models, and 3—area of optimal prediction from the string model.

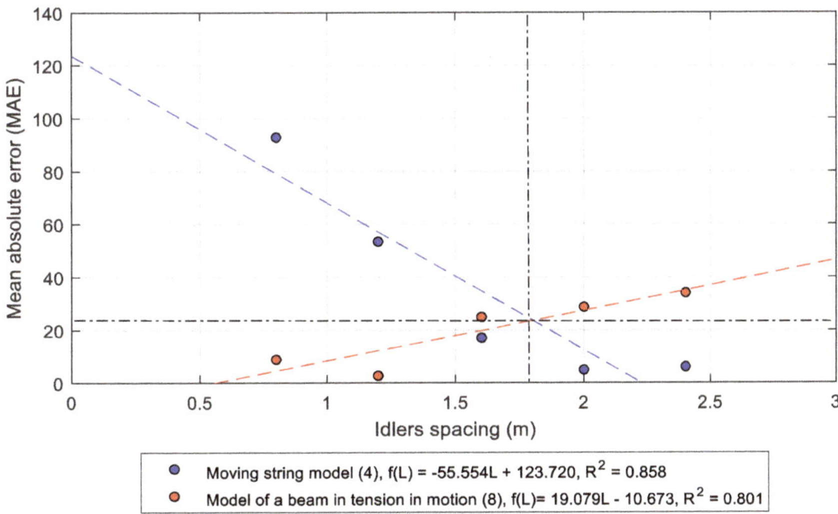

Figure 9. MAE in relation to the string model and to the beam model as a function of idler spacing, with indicated limit idler spacing.

In the case of the analysed ST800 belt, the limit idler spacing for the application of the string model and of the beam model is 1.8 m. For spacing L < 1.8 m, a lower error is observed for the model of beam in tension, while for the spacing L > 1.8 m, the string model offers a better fit. Nevertheless, MAE for the string model is almost two times higher across the entire idler spacing range than MAE for the beam model.

5. Conclusions

Previous research suggests that flexural rigidity in tension members has an influence on the character of their transverse vibrations [37], and this influence was confirmed in the results of measurements here presented. The tests performed on a flat steel-cord belt revealed that its layer-based structure causes it to show a behaviour similar to an elastic beam in tension at small idler support spacing, while when the distances between the supports are greater, the properties of the belt render it similar to an elastic string. The tests focused on a flat belt, as an assumption was made that such a case yields more easily to interpretation at this research stage. Importantly, the conveyor belt in its upper strand is formed in the shape of a trough and is supported by three- or five-idler sets. This fact will most probably contribute to an increased flexural rigidity of the belt, as compared to a flat belt. In the lower strand, the belt can be flat or can be supported on two-idler sets, and therefore, it may have a significantly reduced flexural rigidity as compared to the upper strand. Based on the obtained measurement results and on the typical idler spacing values, the beam model can be assumed to be appropriate for the upper strand (with smaller spacing values), and the string model—for the lower strand (with greater spacing values).

The tests confirmed that the theoretical models properly predicted a significant influence of belt tensile force on the frequencies of transverse vibrations. These changes have a non-linear character, in which the vibration frequency increases together with the force in the belt. Idler spacing is the parameter that affects the similarity between the predictions from the models and the measurement results. The results from the beam model show a smaller error across the entire range of idler spacing values. The model of a tension beam proposed by the authors seems to be a good solution for design calculations. Both the theoretical models and the measurement results indicate a negligible influence of conveyor belt speed on its transverse vibration frequency. The measurement method employed in this research is based on a contactless vibration meter and proved useful in laboratory tests when identifying vibration eigenfrequencies of a moving belt. In the case of idler spacing greater than 1 m, in which the vibration amplitudes are relatively high, the obtained spectra may be strictly interpreted, as the spectra reveal maxima. At small idler spacing, the obtained spectra are not as clear due to lower vibration amplitudes.

The noise and vibration meter allows for clear results only for high vibration amplitudes, and therefore, lower accuracies and less clear spectra are obtained for small idler spacing. The design and the operating principle of the meter itself require the measurements to be performed in certain conditions. Acoustic interferences may cause a complete distortion of the spectrum of the acoustic signal within the low frequency range. This fact may render the presented method impractical in measurements other than those performed in the laboratory. The requirement to locate the measuring head at a close distance above the belt is a further limitation in the case when the belt is loaded with material and the meter cannot be installed in a safe and immovable position. The key focus of further research seems, therefore, to be on finding such a method of belt vibration measurements that would eliminate the above disadvantages of the sound and vibration meter.

Author Contributions: Conceptualization, methodology, software and validation, M.O. and P.B.; writing—review and editing and supervision, R.K. and L.G.; project administration, R.K. and L.G.; final text prepared by M.O., P.B., L.G. and R.K. All authors have read and agreed to the published version of the manuscript.

Funding: The research work was co-founded with the research subsidy of the Polish Ministry of Science and Higher Education granted for 2020.

Institutional Review Board Statement: Not applicable.

Informed Consent Statement: Not applicable.

Data Availability Statement: The data presented in this study are available on request from the corresponding author.

Conflicts of Interest: The authors declare no conflict of interest.

References

1. Suweken, G.; Van Horssen, W.T. On the transversal vibrations of a conveyor belt with a low and time-varying velocity. Part I: The string-like case. *J. Sound Vib.* **2003**, *264*, 117–133. [CrossRef]
2. Gladysiewicz, L. *Belt Conveyors: Theory and Calculations*; Oficyna Wydawnicza Politechniki Wrocławskiej: Wrocław, Poland, 2003; ISBN 83-7085-737-X.
3. Harrison, A. Dynamic behaviour of steel-cord belts. *Colliery Guard.* **1981**, *229*, 459–460.
4. For, O.; Publications, O.; The, O.F.; Communities, E. *Position Paper on Guidelines for the Application of the European Parliament and Council Directive 2000 / 14 / EC on the Approximation of the Laws of the Member States Relating to the Noise Emission in the Environment by Equipment for*; 2000; Volume 162, pp. 1–78.
5. Union, E. Directive 2006/40/EC of the European parliament and of the Council of 17 May 2006, Official J. Eur. Union L 161 (12). *Off. J. Eur. Union* **2006**, *L161*, 12–18.
6. Malinauskaite, J.; Jouhara, H.; Ahmad, L.; Milani, M.; Montorsi, L.; Venturelli, M. Energy efficiency in industry: EU and national policies in Italy and the UK. *Energy* **2019**, *172*, 255–269. [CrossRef]
7. Gładysiewicz, A. Measures of belt conveyor's noise reduction. *Transp. Przem. i Masz. Rob.* **2010**, *3*, 56–63.
8. Idczak, H.; Rudno-Rudzinska, B.; Mazurek, C. [Politechnika W. (Poland)] Reducing noise pollution from belt conveyors in surface mines. *Gor. Odkryw.* **1989**, *31*, 54–69.
9. Sadowski, J.; Fąs, T. Noise minimization of belt conveyor used for lignite coal transportation. *Chem. Eng. Equip.* **2014**, 113–115.
10. Manwar, V.D.; Mandal, B.B.; Pal, A.K. Environmental propagation of noise in mines and nearby villages: A study through noise mapping. *Noise Health* **2016**, *18*, 185–193. [CrossRef] [PubMed]
11. Wickert, J.A.; Mote, C.D. Linear transverse vibration of an axially moving string-particle system. *J. Acoust. Soc. Am.* **1988**, *84*, 963–969. [CrossRef]
12. Pellicano, F.; Vestroni, F. Nonlinear dynamics and bifurcations of an axially moving beam. *J. Vib. Acoust. Trans. ASME* **2000**, *122*, 21–30. [CrossRef]
13. Harrison, A. Determination of the natural frequencies of transverse vibration for conveyor belts with orthotropic properties. *J. Sound Vib.* **1986**, *110*, 483–493. [CrossRef]
14. Lodewijks, G. *Dynamics of Belt Systems, TU Delft*; Delft University of Technology: Delft, The Netherlands, 1996.
15. Suweken, G.; Van Horssen, W.T. On the transversal vibrations of a conveyor belt with a low and time-varying velocity. Part II: The beam-like case. *J. Sound Vib.* **2003**, *267*, 1007–1027. [CrossRef]
16. Andrianov, I.V.; van Horssen, W.T. On the transversal vibrations of a conveyor belt: Applicability of simplified models. *J. Sound Vib.* **2008**, *313*, 822–829. [CrossRef]
17. Ding, H.; Li, D.-P. Static and dynamic behaviors of belt-drive dynamic systems with a one-way clutch. *Nonlinear Dyn.* **2014**, *78*, 1553–1575. [CrossRef]
18. HOU, Y.F.; MENG, Q.R. Dynamic characteristics of conveyor belts. *J. China Univ. Min. Technol.* **2008**, *18*, 629–633. [CrossRef]
19. Abrate, S. Vibrations of belts and belt drives. *Mech. Mach. Theory* **1992**, *27*, 645–659. [CrossRef]
20. Kulinowski, P.; Kasza, P.; Zarzycki, J. The Analysis of Effectiveness of Conveyor Belt Tensioning Systems. *New Trends Prod. Eng.* **2020**, *3*, 283–293. [CrossRef]
21. Manjgo, M.; Piric, E.; Vuherer, T.; Burzic, M. Materials-the Rubber Conveyor Belt With Cartridges Made of Polyester and Polyamide. In Proceedings of the Annals of Faculty Engineering Hunedoara—International Journal of Engineering, Hunedoara, Romania, 2018; pp. 141–145. Available online: http://annals.fih.upt.ro/pdf-full/2018/ANNALS-2018-1-22.pdf (accessed on 1 July 2021).
22. Tokoro, H.; Nakamura, M.; Sugiura, N.; Tani, H.; Yamamoto, K.I.; Shuku, T. Analysis of transverse vibration in engine timing belt. *JSAE Rev.* **1997**, *18*, 33–38. [CrossRef]
23. Ding, H.; Zu, J.W. Effect of one-way clutch on the nonlinear vibration of belt-drive systems with a continuous belt model. *J. Sound Vib.* **2013**, *332*, 6472–6487. [CrossRef]
24. Beikmann, R.S.; Perkins, N.C.; Ulsoy, A.G. Free vibration of serpentine belt drive systems. *J. Vib. Acoust. Trans. ASME* **1996**, *118*, 406–413. [CrossRef]
25. Scurtu, P.R.; Clark, M.; Zu, J.W. Coupled longitudinal and transverse vibration of automotive belts under longitudinal excitations using analog equation method. *JVC J. Vib. Control* **2012**, *18*, 1336–1352. [CrossRef]
26. Moon, J.; Wickert, J.A. Non-linear vibration of power transmission belts. *J. Sound Vib.* **1997**, *200*, 419–431. [CrossRef]
27. Hedrih, K. Transversal vibrations of the axially moving sandwich double belt system with creep layer. *IFAC Proc. Vol.* **2006**, *39*, 167–172. [CrossRef]
28. Kim, S.K.; Lee, J.M. Analysis of the non-linear vibration characteristics of a belt-driven system. *J. Sound Vib.* **1999**, *223*, 723–740. [CrossRef]
29. Pukach, P.; Sokhan, P.; Stolyarchuk, R. Investigation of mathematical models for vibrations of one dimensional environments with considering nonlinear resistance forces. *ECONTECHMOD An Int. Q. J. Econ. Technol. Model. Process.* **2016**, *5*, 97–102.
30. Nikolaevna Aleksandrova, T. Interaction of lunch ore with conveyor belt. The original in Russian: "Взаимодействие кусковой руды с лентой конвейера". *Eastern-European J. Enterp. Technol.* **2008**, *35*, 48–52.

31. Pihnastyi, O.; Khodusov, V.; Kozhevnikov, G.; Bondarenko, T. Analysis of Dynamic Mechanic Belt Stresses of the Magistral Conveyor. In *Proceedings of the Grabchenko's International Conference on Advanced Manufacturing Processes*; Springer: Cham, Switzerland, 2021; pp. 186–195.

32. Lodewijks, G. Two Decades Dynamics of Belt Conveyor Systems. 2007. Available online: https://login.totalweblite.com/Clients/doublearrow/beltcon%202001/1.two%20decades%20dynamics%20of%20belt%20conveyor%20systems.pdf (accessed on 1 July 2021).

33. Harrison, A. Criteria for minimising transient stress in conveyor. *Trans. Inst. Eng., Aust., Mech. Eng.* **1983**, *ME8*, 129–134.

34. Kong, L.; Parker, R.G. Coupled belt-pulley vibration in serpentine drives with belt bending stiffness. *J. Appl. Mech. Trans. ASME* **2004**, *71*, 109–119. [CrossRef]

35. Ravindra, V.; Padmanabhan, C.; Sujatha, C. Static and free vibration studies on a pulley-belt system with ground stiffness. *J. Brazilian Soc. Mech. Sci. Eng.* **2010**, *32*, 61–70. [CrossRef]

36. Hop, T. Vibrations of compressed beams. The original in Polish: "Drgania belek sprężonych," Politechnika Śląska. 1962. Available online: http://delibra.bg.polsl.pl/Content/2513/Hop.pdf (accessed on 1 July 2021).

37. American Society of Mechanical Engineers. Design Engineering Div. Power Transmission and Gearing. In Proceedings of the 1989 International Power Transmission and Gearing Conference, Chicago, IL, USA, 25–28 April 1989; pp. 25–29.

38. Wasfy, T.M.; Leamy, M. Effect of bending stiffness on the dynamic and steady-state responses of belt-drives. In Proceedings of the International Design Engineering Technical Conferences and Computers and Information in Engineering Conference, Montreal, QB, Canada, 1 January 2002; Volume 36533, pp. 217–224.

39. Callegari, M.; Cannella, F.; Ferri, G. Multi-body modelling of timing belt dynamics. *Proc. Inst. Mech. Eng. Part K J. Multi-body Dyn.* **2003**, *217*, 63–75. [CrossRef]

40. Ulsoy, A.G.; Whitesell, J.E.; Hooven, M.D. Design of belt-tensioner systems for dynamic stability. *J. Vib. Acoust. Trans. ASME* **1985**, *107*, 282–290. [CrossRef]

41. Hu, Y.; Yan, Y.; Wang, L.; Qian, X. Non-Contact Vibration Monitoring of Power Transmission Belts Through Electrostatic Sensing. *IEEE Sens. J.* **2016**, *16*, 3541–3550. [CrossRef]

42. Ma, X.; Shi, X.; Zhang, J. Modeling and experimental investigation on the vibration of main drive chain in escalator. In Proceedings of the INTER-NOISE 2019 MADRID—48th International Congress and Exhibition on Noise Control Engineering, InterNoise19, Madrid, Spain, 17–19 June 2019; Institute of Noise Control Engineering: Indianapolis, IN, USA, 2019. Volume 259. pp. 2069–2080.

43. Bortnowski, P.; Gladysiewicz, L.; Krol, R.; Ozdoba, M. Tests of belt linear speed for identification of frictional contact phenomena. *Sensors* **2020**, *20*, 5816. [CrossRef] [PubMed]

44. Gładysiewicz, L.; Kawalec, W.; Król, R. Selection of carry idlers spacing of belt conveyor taking into account random stream of transported bulk material. *Maint. Reliab.* **2016**, *18*, 32–37. [CrossRef]

Article

Simple Design Solution for Harsh Operating Conditions: Redesign of Conveyor Transfer Station with Reverse Engineering and DEM Simulations

Błażej Doroszuk *, Robert Król and Jarosław Wajs

Faculty of Geoengineering, Mining and Geology, Wroclaw University of Science and Technology, Na Grobli 15, 50-421 Wrocław, Poland; robert.krol@pwr.edu.pl (R.K.); jaroslaw.wajs@pwr.edu.pl (J.W.)
* Correspondence: blazej.doroszuk@pwr.edu.pl

Abstract: This paper addresses the problem of conveyor transfer station design in harsh operating conditions, aiming to identify and eliminate a failure phenomenon which interrupts aggregate supply. The analyzed transfer station is located in a Polish granite quarry. The study employs laser scanning and reverse engineering methods to map the existing transfer station and its geometry. Next, a discrete element method (DEM) model of granite aggregate has been created and used for simulating current operating conditions. The arch formation has been identified as the main reason for breakdowns. Alternative design solutions for transfer stations were tested in DEM simulations. The most uncomplicated design for manufacturing incorporated an impact plate, and a straight chute has been selected as the best solution. The study also involved identifying areas of the new station most exposed to wear phenomena. A new transfer point was implemented in the quarry and resolved the problem of blockages.

Keywords: conveyor belts; transfer station; DEM simulation; reverse engineering

Citation: Doroszuk, B.; Król, R.; Wajs, J. Simple Design Solution for Harsh Operating Conditions: Redesign of Conveyor Transfer Station with Reverse Engineering and DEM Simulations. *Energies* **2021**, *14*, 4008. https://doi.org/10.3390/en14134008

Academic Editors: Daniela Marasova, Monika Hardygora and Mirosław Bajda

Received: 25 May 2021
Accepted: 28 June 2021
Published: 2 July 2021

1. Introduction

Transfer points are critical components of large belt conveyor systems in the mining industry, and their proper functioning is essential for the proper performance of other elements. Breakdowns caused by transfer point failures contribute to enormous energy losses due to downtime in the supply of the transported material and to sub-optimal use of available technological resources [1]. Improper design of the transfer station also causes increased resistance at the loading point of the conveyor belt [2], thus increasing the energy demand of the drive system for a particular conveyor. When analyzing the energy consumption of the belt conveyor system as a whole, the impact of transfer station design and its optimization cannot be ignored. An optimally designed transfer point also reduces potential damage to the receiving belt caused by falling solids if these are of a size that can damage the conveyor belt [3]. Researchers have extensively explored the topic of the potential damage to the belt from falling objects [4–6]. Therefore, a properly designed station decreases the costs of conveyor belt replacement.

Many previous generation industrial plants, including mines, have transfer points that were designed following a rule of thumb [7]. They often do not have detailed documentation, and after each failure, they were modified by trial and error. Additional metal sheets are often welded to the structure in order to prevent the material from spilling or to solidify the areas most exposed to abrasion. On occasion, some elements are cut to make more space for the material stream after a blockage. Poorly designed transfer points are often repaired using only makeshift measures, and after a critical failure, the entire construction has to be redesigned. The best solution would be to replace the old elements entirely with new versions, designed according to the latest standards. However, the complete removal of a structure and the construction of a new one is a very expensive task, and not every plant can afford such an undertaking. Therefore, it is economically justified to cut out only some of the unnecessary elements of the old transfer station and fit new ones into the existing structure. Reverse engineering methods prove very useful for designs that do not have sufficient documentation. These methods allow mapping the geometry of the current equipment digitally, in the form of CAD models. Engineers can easily modify such models and validate them with simulation methods.

For years, continuum-based analytical methods were the only tools available to transfer station designers [8]. Several studies have made comparisons between continuum-based analysis and computational methods [8,9]. When designing transfer stations operating in difficult conditions, the most frequently used method is the discrete element method (DEM), one of the most important simulation techniques developed for the analysis of particulate systems [10]. The method has gained popularity with the increase of computing power, and the first industrial applications date back to the early 1990s [11]. In 1994, Nordell successfully predicted that engineers would extensively use this method for transfer station design [12]. DEM enables simulations of the behavior of bulk materials. The simulations are on the level of separate material particles and their interactions with each other and with the geometry understood as boundaries for particle movement. With this method, it is possible to map specific conditions under which failures occur in existing transfer points and conduct an in-depth analysis in order to understand the mechanism behind, e.g., a blockage in the transfer point.

This paper will discuss an optimization process performed for an existing transfer station in a quarry. The discharge connects two conveyors at an acute angle, where the first conveyor is wider than the second conveyor, with the latter additionally being sharply sloped. Further, in the current situation and under specific conditions, the risk of blockage highly increases. The latest works on DEM application in designing and analyzing the efficiency of transfer stations refer to stations with a known geometry [9,13,14]. A novelty in this work is the combination of laser scanning to map an existing transfer point and then the use of real geometry in DEM simulations for further analysis.

2. Case Study

The optimized transfer station is located in the "Graniczna" quarry in the Dolnośląskie Voivodeship in Poland. The quarry extracts granite and processes it into aggregate. The discharge connects two belt conveyors, WD-1 (feeding) (Figure 1a) and B-1 (receiving) (Figure 1b).

(a)   (b)

Figure 1. Conveyors connected by analyzed transfer station: (**a**) feeding conveyor belt WD-1 and (**b**) receiving conveyor belt B-1.

The WD-1 conveyor has a 1.2 m wide belt (Cobra DX FLEXAMID 3150/1 1200 10 + 4), is inclined at an angle of 7.39°, and has a velocity of 1.63 m/s, while the B-1 conveyor has a 1 m wide belt (Wolbrom 1000 EP 800 4 4 + 2) inclined at an angle of 16.4°, and has the velocity of 1.85 m/s. The DX FLEXAMID belt is a 1-ply belt with textile carcass and rubber covers, and the Wolbrom belt is a 4-ply polyamide-polyester belt. The temporary value of the capacity recorded on the crusher feeding the WD-1 reaches almost 900 Mg/h (Figure 2). The crushed product fed to conveyor WD-1 has grain composition, as presented in Figure 3.

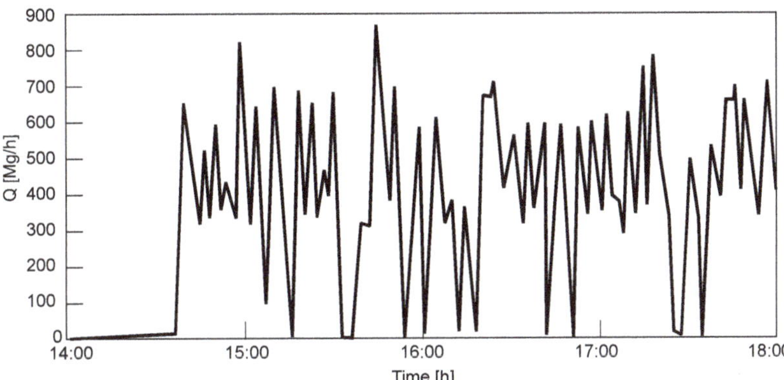

Figure 2. Capacity of the crusher feeding the WD-1 conveyor.

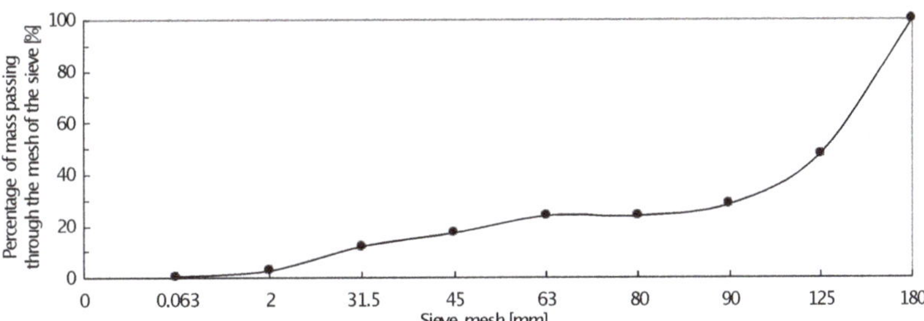

Figure 3. Cumulative particle size distribution.

The angle between the conveyors is 23.6°, and the height difference is about 2 m. It means that transfer station WD-1/B-1 has to change the direction, in a very limited space, of the aggregate stream by 156.4° in the horizontal plane.

The current transfer station is hard to classify as a specific type of construction, but it has some "rock box" features. There are shelves inside (Figure 4) that allow for material retention. The stream of granite does not directly hit the elements of the station, it only loses energy on the piled aggregate. Then, the material is discharged onto the receiving conveyor. The construction has been modified many times, as evidenced by numerous irregular welded plates and a cut-out in the front wall, increasing the space for the aggregate stream. However, despite the modifications, blockages still occur under unfavorable working conditions, i.e., during bad weather, at very high temporary capacity, and also in the case of larger rocks.

The requirement for the modification process was to create a simple-to-build transfer point that would fit into the current main structure of the transfer point. The engineers in the quarry expressed a preference for flat surfaces to be included in the new design.

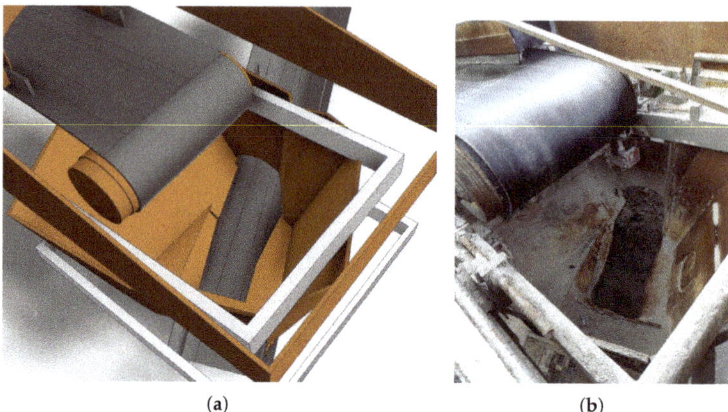

(a)   (b)

Figure 4. The inside of the transfer station: (**a**) CAD model and (**b**) existing station.

3. Materials and Methods

3.1. Laser Scanning

The literature shows that terrestrial laser scanning is commonly applied in mining areas and allows rapid data acquisition for developing 3D mine models [15], ensuring slope stability [16], and detecting changes based on two TLS point clouds [17]. The application of TLS data, as presented in [18], proves the usefulness of 3D point cloud data for structural health monitoring (SHM) of constructed objects.

LiDAR data from a laser scanner allow the monitoring of long-range engineering objects (e.g., a steel trestle bridge) [19] and the modeling of mining shaft deformations [20]. The current status of the development trends in, and the potential for, TLS application in mining areas were described in detail in [21]. Based on active sensors, LiDAR techniques help to detect the shape of the object, which can be transported by the conveyor belt [22] and detect the damage [23]. In the present case study, the acquired point cloud allowed the building of a 3D BIM model in order to reconstruct the shape of the analyzed equipment (transfer station).

The LiDAR data were acquired following the terrestrial approach (TLS) with the multi-station acquisition, significantly reducing the number of potential occlusions. The 3D point cloud was obtained from the Riegl V400i terrestrial laser scanner (Figure 5a). The point cloud post-processing involved full multi-station adjustment with high internal and external quality of the geo-referenced final point cloud. The resulting post-processed scans represent the full metric point cloud, which can be used to identify, design, and measure objects in a metric 3D model (Figure 5b). The product of the 3D laser scanning procedure provides excellent data for the 3D modeling of mining objects in the BIM (building information modeling) technology. The final point cloud provides information about the intensity, amplitude, and, optionally, RGB values for each point in the point cloud. TLS measurements allow the reverse engineering method to be employed for future CAD modeling of the analyzed transfer stations based on the original geometry from the TLS point cloud.

3.2. Theoretical Capacity of the Conveyor Belts

Material flow through curved or straight chutes can be categorized as "fast" or "slow" [24]. When the chute works under fast flow conditions, the material stream thickness varies along the chute, with minimum thickness near the area of maximum stream velocity [24]. Solutions providing fast flow conditions at the maximum capacity of the feeding belt will be tested as part of further research works. Belt capacity should correspond to peaks of material volumetric discharge from the feeding point [25]. In the analyzed case, the material supplied by trucks and loaders to the jaw crusher is fed to the conveyor. The nominal capacity of the conveyors was determined from theoretical cross-sections and from belt cross-sections obtained in the scanning procedure (Figure 6). The theoretical cross-section was calculated from the dependencies described in [26]. The width of the supplied material was projected from the theoretical cross-sections onto the actual cross-sections of the belts, and the material surface area was drawn using the same angles of repose as above.

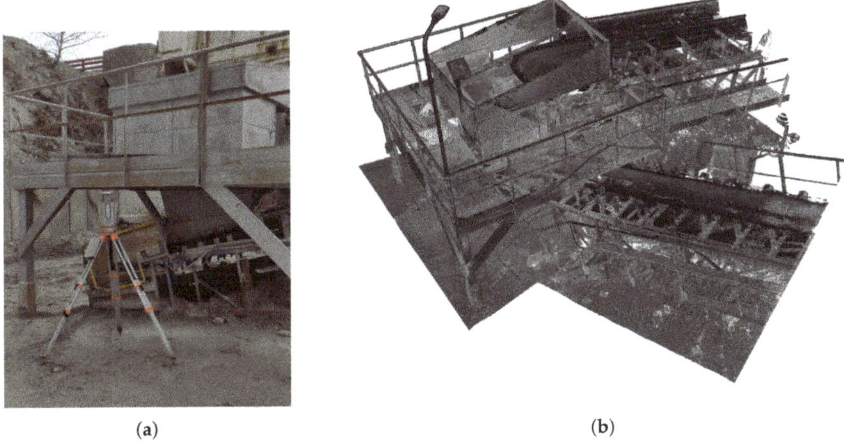

(a)   (b)

Figure 5. Laser scanning: (**a**) Riegl V400i scanner and (**b**) point cloud of transfer station.

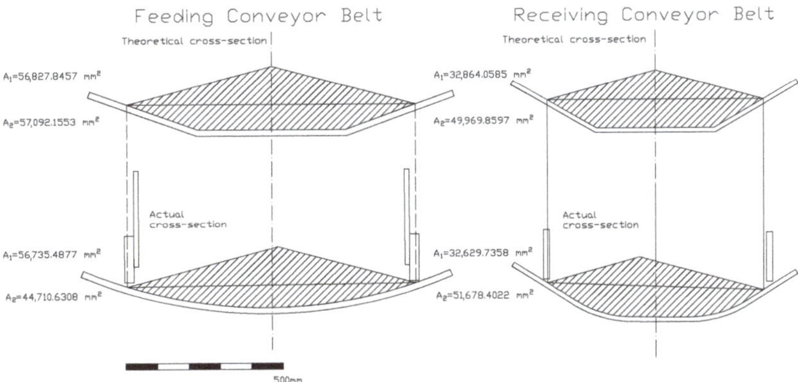

Figure 6. Cross-section of loaded conveyor belts.

It is worth noting that when scanned, the belts were empty, so the actual capacity of the conveyors will not be less than the capacity determined for the existing cross-sections of the empty belt. Still, it should not be greater than the nominal capacity determined analytically. As a result, the capacity remains within the following ranges:

- WD-1: 837–1123 Mg/s
- B-1: 692–751 Mg/s

Conveyor B-1 has a nominal capacity smaller than the maximum capacity of the crusher that feeds conveyor WD-1, which may lead to problems.

3.3. CAD Model

Using CAD tools, the solids were adjusted to the point cloud. The mapping was limited to only those elements of the geometry of the transfer station which may come into direct contact with the bulk material and the elements limiting the construction space.

Knowledge of space limitations is necessary to fit a new transfer station into the existing main construction. The geometry has been simplified and is represented by flat faces (Figure 7).

Figure 7. CAD model of the transfer station.

3.4. DEM Model

The importance of selecting proper model parameters was demonstrated in [27]. In the publication, granite aggregate samples (16–22 mm) and several larger granite rocks were collected from a quarry. Three cylindrical specimens with a diameter of 43 mm and a height of 86 mm were cut from the larger rocks, which were then tested by the uniaxial compression test. The tested granite had an average density of 2612 kg/m^3, a Poisson's ratio of 0.11, and a Young's modulus of 18.4 GPa. The friction coefficient of granite against steel and the conveyor belt rubber was also investigated using an inclined plane, obtaining average values of 0.46 and 0.76, respectively.

The aggregate particles were mapped as four conglomerates of spheres, consisting of 4 to 7 spheres to reflect the irregular shape of the particles. Parameters that cannot be measured directly were calibrated by imitating the material flow in the calibration box (Figure 8a). The calibration aimed to achieve a set of DEM parameters that provide a simulation of the flowing material in approximation to the actual conditions with respect to the time flow and the granite angle of repose. The edge of the piled-up aggregate was redrawn on the photo manually, and then the trend line was matched with the dots, as shown in the figure below (Figure 8b). The calibration process was similar to that in the previous works [28,29]. Researchers [30] have shown that obtaining correspondence in two different parameters during an experimental simulation may be sufficient to obtain a unique set of parameters, allowing the behavior of bulk material to be adequately simulated. The obtained parameters from calibration are presented in Table 1 and the material parameters used during calibration and simulations are presented in Table 2. Young's modulus of the Granite was reduced 2000 times, to 9.2 MPa, so the simulation time-step time dependent on that parameter could be longer. The value of Young's modulus was chosen as minimal as possible, which provided the same results in simulations as the original but drastically reduced the computing time. Simulations were performed with the time-step equal to 0.00024 s.

Table 1. Calibrated parameters.

Coefficient of	Granite-Granite	Granite-Rubber	Granite-Steel
Restitution	0.2	0.3	0.2
Static Friction	0.7	0.53	0.32
Rolling Friction	0.05	0.01	0.01

Table 2. Material parameters.

Parameter	Granite	Rubber	Steel
Density (kg/m^3)	2610	1100	7800
Young's Modulus (MPa)	18,400 (9.2)	8.69	200,000
Poisson's Ratio (-)	0.11	0.499	0.3

(a)   (b)

Figure 8. Calibration: (**a**) scheme of the calibration box and (**b**) granite angle of repose.

The grain composition of the simulated material was adjusted (based on data shown in Figure 3) using normal distribution, with an average of 110 mm and a standard deviation of 40 mm, so that the mean grains of the simulated and of the actual material were similar. Solids smaller than 50 mm were rejected in the simulation.

3.5. DEM Simulations

The simulations were carried out for normal (300 Mg/h) and extreme conditions (900 Mg/h), as well as for humidity causing reduced friction coefficients and for increased number of solids of up to 300 mm in diameter. During one time step, the maximum generated amount of particles was approximately 1500, which corresponds to about 9000 total spheres building all particles. Due to a small percentage of the dust fraction, cohesive interactions were neglected. Research shows that rubber friction for wet surfaces is 20–30% smaller than for corresponding dry surfaces [31]. A 30% reduction in particle-geometry friction was assumed when humidity was taken into account for further analysis. The first simulations were carried out for a transfer station according to the original design, and subsequent simulations for alternative considered solutions. In the EDEM Software simulation environment, the motion called "conveyor translation" was assigned to conveyor belt geometries. Conveyor translation makes particles in the contact of the belt surface move with the belt velocity and direction. The belt in the simulation does not move, but bulk material that comes into interaction with the belt surface moves like on the real conveyor belt.

4. Results of Numerical Analyses

4.1. Mapping the Work of the Transfer Station

The performance of the transfer station according to the original design was mapped for the simulated grain composition and for a capacity of 300 Mg/h. No disturbing phenomena were observed in the case of these performance settings. The material was placed on a shelf inside the station of the original solution, and then it fell onto the receiving belt.

Several simulations were performed in order to identify the failure mechanism. They considered factors such as the capacity increased to 900 Mg/h (Figure 9a), a greater number of solids having a diameter of up to 300 mm, and increased humidity (lower friction coefficients against steel and rubber).

It has been observed that the performance increase itself does not cause failures. However, with other factors included, the phenomenon of "arch" formation was observed (Figure 9b) in the transfer station discharge opening. When larger particles account for 30 percent of the excavated material, the arch may be formed at a temporary capacity of 800 Mg/h. However, with moisture and large solids, the phenomenon can be observed even at a temporary capacity of 600 Mg/h.

(**a**) (**b**)

Figure 9. Observed failure: (**a**) simulation of material flow through the transfer station at a capacity of 900 Mg/h and (**b**) the observed arch formation.

The discharge opening of the current transfer station is similar to the discharge of the hopper. In the case of the hopper, the space for mass flow must be sufficient to prevent the formation of the cohesive/mechanical arch. Research [32] suggests that the opening width should be at least four times the maximum particle dimension. Under the analyzed conditions, this would mean that the opening should be 1200 mm wide, i.e., 200 mm more than the width of the receiving conveyor belt.

4.2. Alternative Designs

Possible modifications to the present transfer station were tested under the worst-case conditions. A minimal improvement was achieved by designing a new rock box, but it did not meet the requirements (it was not failure-free in the worst analyzed conditions). Such a design solution does not change the method behind the transfer station discharge, which is still similar to unloading the hopper, the mass-flow opening of which should be wider, as mentioned previously. Modifications of the existing geometry were limited by the main steel frame of the transfer station and its location in the transportation system.

However, numerous research works have already demonstrated a potential for the use of curved elements in transfer stations. They provide an easy solution for redirecting the stream of material [2]. Curved chutes have many advantages: (a) reducing the need for belt covers of a significant thickness, (b) no need for large steel cord filament diameters to resist impacts, and (c) spillage prevention [12]. Straight chutes are easier to manufacture and have favorable material streamflow characteristics [24].

Hence, the idea of the hood and spoon transfer operation was tested. This type of design provided satisfactory results. The station consists of two profiled elements redirecting the stream of material in the desired direction. However, it is a complicated structure that requires bent parts and is currently not considered by the plant staff for the new design. The main advantage of the construction is that it feeds the aggregate stream continuously, and it does not accumulate any material that could later cause blockage, as is the case with the present construction. Therefore, it was decided to use the idea of continuous feeding with the most straightforward possible structure. A satisfactory effect was obtained using a structure consisting of two main steel elements: an inclined impact plate, which redirects the stream of material downwards, and then a straight chute, along which the granite slides onto the receiving belt (Figure 10a). As shown in Figure 10b, the area in which the material accelerates to reach the belt speed is very short.

(**a**) (**b**)

Figure 10. Simulation of material flow through the new transfer station at a capacity of 900 Mg/h: (**a**) visualization and (**b**) velocity of the particles.

The current station has been compared with the design, in which the steel impact plate is 770 mm from the feeding conveyor head pulley axis, and the chute is set at an angle of 60°. The influence of the impact plate inclination by 10° and a change in the inclination of the bottom part of the chute to 45° (an angle slightly greater than the angle of repose of natural granite) were also investigated. The impact plate was placed in such a position that the material was redirected into the center of the chute. The angles of the chute were limited by the available space.

Comparisons also involved the distribution of cumulative normal energy lost by the material stream on the belt during collisions. The surface that receives impacts in the current station is concentrated in the center of the belt, which may cause uneven wear (Figure 11).

The total normal energy that the belt acquires during the simulation for each of the analyzed variants of the transfer station is shown in Figure 12. The total normal energy was exported for the successive time steps of the simulation, with a constant time interval. Then, using linear regression, the trend lines were fitted to the data. It shows that by using a 45° chute, it is possible to reduce the energy that the material loses on the receiving belt by more than 10 percent compared to the currently lost energy.

The design of an angular discharge is obligatory so that the material is evenly fed on the belt, as an uneven placement of the aggregate stream may cause misalignment [33]. The use of the chute ensures even distribution of the material along the width of the belt.

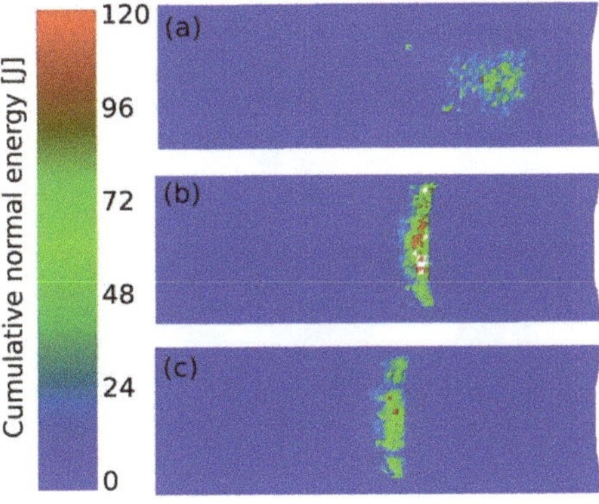

Figure 11. Distribution of the cumulative normal energy transferred by the material stream to the receiving belt after 60 s of simulation: (**a**) current transfer station, (**b**) 60° chute, and (**c**) 45° chute.

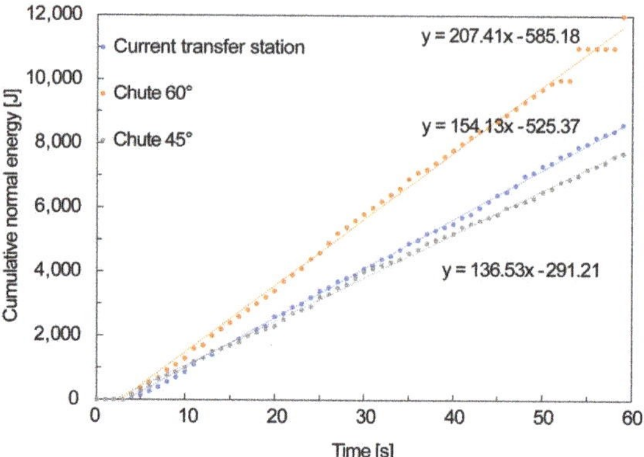

Figure 12. Total cumulative normal energy absorbed by the belt during collisions with the material for a capacity of 300 Mg/h.

4.3. Determination of the Elements Most Exposed to Wear

In addition to the energy loss due to collisions with the receiving belt, the material also loses energy as it slides down the chute. Studies have shown that under thin-stream accelerated flow, the energy losses are approximately: 82% due to sliding along the chute bottom, 9% due to sliding against sidewalls, and 9% due to inter-granular friction [34].

During the simulation, it is possible to analyze many different parameters resulting from the interaction of the bulk material with the elements of the geometry. The energy transferred to the geometry depends on the mass and velocity of the particles during the collisions. The tangential and normal energies that act on the impact plate were investigated when this was mounted vertically and inclined by 10°, and on the chute in the variant in which the entire chute was inclined at an angle of 60° and in which the bottom part of the chute was inclined at an angle of 45°. The inclination of the impact plate by 10° was observed to have a positive effect on reducing the normal energy loss by the impacting material stream, which helps to reduce the noise and the damage to the element. The tilt also causes a more even distribution of the tangential energy (Figure 13) (the energy responsible for the abrasions on the surface). The 45° inclination angle of the chute can provide more even distribution of the tangential energy (Figure 14).

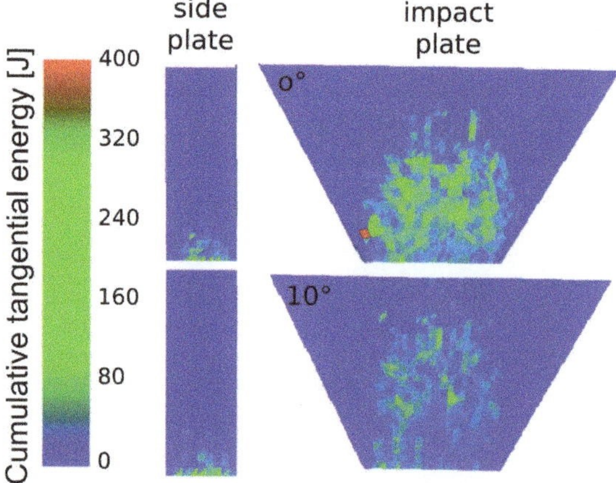

Figure 13. Cumulated tangential energy after 60 s of simulation for the impact plate and the side plate mounted vertically and at an angle of 10° of the structure, for 900 Mg/h.

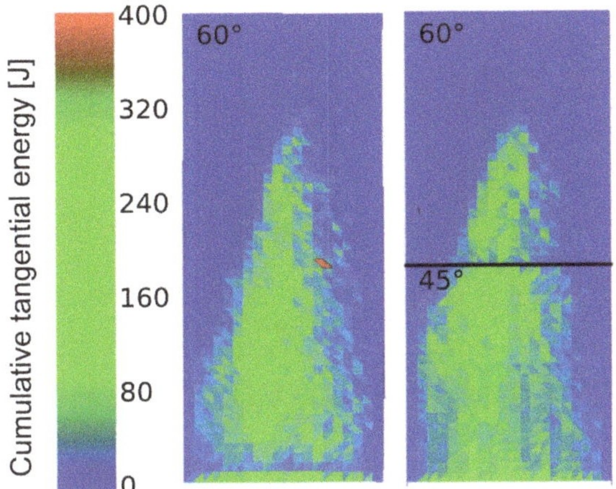

Figure 14. Cumulated tangential energy after 60 s of simulation on the chute, for 900 Mg/h.

4.4. Implementation of the New Design

The variant with an inclined impact plate and a chute with two parts of different inclination was chosen as the best and simplest solution. This design of the transfer station has been implemented (Figure 15a) in the mining plant and successfully fulfills its task (Figure 15b). The elements of the transfer station were reinforced with abrasion-resistant steel, which will ensure the long life of these elements. The elimination of the risk of blockages and the positioning of the spoil stream centrally on the receiving conveyor demonstrate the suitability of the granite DEM model for optimization activities.

(**a**) (**b**)

Figure 15. New transfer station: (**a**) prepared, and (**b**) in operation.

5. Discussion and Conclusions

For the purpose of this research, a laser scan was performed of the existing transfer station in the "Graniczna" mine, between the WD-1 and the B-1 conveyors. A CAD model for the DEM simulation was developed from the obtained point cloud using reverse engineering methods. Nominal capacities of the conveyors were determined with allowance for their parameters. Examinations included the material parameters of granite and the coefficients of friction against steel and rubber. The remaining parameters were calibrated against the flow time and the angle of repose in the calibration box. Then, the normal operating conditions of the transfer station were mapped in the simulation, subsequently involving the potential unfavorable conditions. The results of the simulations allowed the design defects of the current construction to be determined.

An alternative transfer station in the form of an impact plate and a chute was proposed. The energy lost by the aggregate stream on the belt and on the transfer station elements was also determined for the proposed variants. Cumulative energy lost on the chute and on the belt by the material stream was chosen as an optimization criterion. The quarry staff implemented the proposed design solution in the analyzed transfer point. The transfer station fulfilled the tasks set for it and contributed to reducing the failure rate of the conveyor belt.

The most important conclusions from the conducted analysis are as follows:

- The receiving conveyor is inclined at a large angle, and the temporary capacities exceed its theoretical nominal capacity, which may lead to unfavorable phenomena. Regardless of the transfer station design, if the material stream capacity is greater than the nominal capacity of the receiving conveyor, a failure may occur.
- The previous design of the station was sufficient when the material was dry, and the grain distribution was in accordance with the one presented in Figure 3.
- When unfavorable factors coincide, i.e., the percentage of large solids increases, the temporary capacity increases in wet conditions, and an arch may be formed, blocking the discharge space at a temporary capacity of approximately 600 Mg/h.
- The functionality of the transfer station was improved by installing elements that ensure continuous transport of the material without accumulating the material on the shelves.
- The station, which consists of an impact plate and a chute, allows for a failure-free transfer of aggregate material in difficult operating conditions.
- The use of a chute installed at an angle of 45° reduced the energy lost by the material on the belt by more than 10 percent.
- The impact plate, inclined at an angle of 10°, is characterized by lower wear than the vertical variant.

Author Contributions: Conceptualization, B.D. and R.K.; methodology, B.D.; software, B.D. and J.W.; validation, B.D.; formal analysis, B.D.; investigation, B.D.; resources, B.D., J.W. and R.K.; data curation, B.D. and J.W.; writing—original draft preparation, B.D.; writing—review and editing, R.K. and J.W.; visualization, B.D.; supervision, R.K.; project administration, R.K.; funding acquisition, R.K. All authors have read and agreed to the published version of the manuscript.

Funding: The research work was co-funded with the research subsidy of the Polish Ministry of Science and Higher Education granted for 2021.

Institutional Review Board Statement: Not applicable.

Informed Consent Statement: Not applicable.

Data Availability Statement: The data presented in this study are available on request from the corresponding author.

Acknowledgments: We want to acknowledge cooperation with the "Graniczna" quarry from EUROVIA KRUSZYWA S.A. The quarry management provided us with objects to analyze and handled the project implementation and tests.

Conflicts of Interest: The authors declare no conflict of interest.

References

1. Czuba, W.; Furmanik, K. Analiza ruchu ziarna w przestrzeni przesypowej przenośnika taśmowego. *Eksploat. I Niezawodn.* **2013**, *15*, 390–396.
2. Roberts, A.; Arnold, P. Discharge—Chute design for free—Flowing granular materials. *Trans. Am. Soc. Agric. Eng.* **1971**, *14*. [CrossRef]
3. Rudolf, L.; Durna, A.; Hapla, T. The issue of the transfer points on belt conveyors. *Int. Multidiscip. Sci. GeoConf. Surv. Geol. Min. Ecol. Manag. SGEM* **2018**, *18*, 989–996. [CrossRef]
4. Grincova, A.; Andrejiova, M.; Marasova, D.; Khouri, S. Measurement and determination of the absorbed impact energy for conveyor belts of various structures under impact loading. *Meas. J. Int. Meas. Confed.* **2019**, *131*, 362–371. [CrossRef]

5. Fedorko, G.; Molnar, V.; Marasova, D.; Grincova, A.; Dovica, M.; Zivcak, J.; Toth, T.; Husakova, N. Failure analysis of belt conveyor damage caused by the falling material. Part I: Experimental measurements and regression models. *Eng. Fail. Anal.* **2014**, *36*, 30–38. [CrossRef]
6. Ľubomír, A.; Taraba, V.; Szabo, S.; Leco, M. Belt damage aspect to impact loading. *Appl. Mech. Mater.* **2014**, *683*, 102–107. [CrossRef]
7. Gutierrez, A.; Garate, G. Design of a transfer chute for multiple operating conditions. *ASME Int. Mech. Eng. Congr. Expo. Proc. (IMECE)* **2014**, *11*. [CrossRef]
8. Hastie, D.B.; Wypych, P.W. Experimental validation of particle flow through conveyor transfer hoods via continuum and discrete element methods. *Mech. Mater.* **2010**, *42*, 383–394. [CrossRef]
9. Ilic, D.; Roberts, A.; Wheeler, C.; Katterfeld, A. Modelling bulk solid flow interactions in transfer chutes: Shearing flow. *Powder Technol.* **2019**, *354*, 30–44. [CrossRef]
10. Katterfeld, A.; Wensrich, C. Understanding granular media: From fundamentals and simulations to industrial application. *Granul. Matter* **2017**, *19*. [CrossRef]
11. Gröger, T.; Katterfeld, A. Application of the discrete element method in materials handling—Part 3: Transfer stations. *Bulk Solids Handl.* **2007**, *27*, 158–166.
12. Nordell, L.K. Palabora installs curved transfer chute in hard rock to minimize belt cover wear. *Bulk Solids Handl.* **1994**, *14*, 739–743.
13. Rossow, J.; Coetzee, C. Discrete element modelling of a chevron patterned conveyor belt and a transfer chute. *Powder Technol.* **2021**, *391*, 77–96. [CrossRef]
14. Ilic, D.; Lavrinec, A.; Orozovic, O. Simulation and analysis of blending in a conveyor transfer system. *Miner. Eng.* **2020**, *157*, 106575. [CrossRef]
15. Xu, Z.; Xu, E.; Wu, L.; Liu, S.; Mao, Y. Registration of terrestrial laser scanning surveys using terrain-invariant regions for measuring exploitative volumes over open-pit mines. *Remote Sens.* **2019**, *11*, 606. [CrossRef]
16. Bazarnik, M. Slope stability monitoring in open pit mines using 3D terrestrial laser scanning. In *E3S Web of Conferences*; EDP Sciences: Julis, France, 2018; Volume 66, p. 01020.
17. Vassena, G.; Clerici, A. Open pit mine 3D mapping by tls and digital photogrammetry: 3D model update thanks to a slam based approach. *Int. Arch. Photogramm. Remote Sens. Spat. Inf. Sci.* **2018**, *42*, 1145–1148. [CrossRef]
18. Yang, H.; Xu, X.; Neumann, I. Laser scanning-based updating of a finite-element model for structural health monitoring. *IEEE Sens. J.* **2015**, *16*, 2100–2104. [CrossRef]
19. Skoczylas, A.; Kamoda, J.; Zaczek-Peplinska, J. Geodetic monitoring (TLS) of a steel transport trestle bridge located in an active mining exploitation site. *Ann. Wars. Univ. Life Sci. Sggw. Land Reclam.* **2016**, *48*, 255–266. [CrossRef]
20. Lipecki, T.; Kim, T.T.H. The Development of Terrestrial Laser Scanning Technology and Its Applications in Mine Shafts in Poland. *Inżynieria Miner.* **2020**, *1*, 301–310.
21. Fengyun, G.; Hongquan, X. Status and development trend of 3D laser scanning technology in the mining field. In *2013 the International Conference on Remote Sensing, Environment and Transportation Engineering (RSETE 2013)*; Atlantis Press: Nanjing, China, 2013.
22. Fojtík, D. Measurement of the volume of material on the conveyor belt measuring of the volume of wood chips during transport on the conveyor belt using a laser scanning. In Proceedings of the 2014 15th International Carpathian Control Conference (ICCC), Velke Karlovice, Czech Republic, 28–30 May 2014; pp. 121–124.
23. Trybała, P.; Blachowski, J.; Błażej, R.; Zimroz, R. Damage Detection Based on 3D Point Cloud Data Processing from Laser Scanning of Conveyor Belt Surface. *Remote Sens.* **2021**, *13*, 55. [CrossRef]
24. Roberts, A.W. An Investigation of the Gravity Flow of Noncohesive Granular Materials Through Discharge Chutes. *J. Manuf. Sci. Eng. Trans. ASME* **1969**, *91*, 373–381. [CrossRef]
25. Kruczek, P.; Polak, M.; Wyłomańska, A.; Kawalec, W.; Zimroz, R. Application of compound Poisson process for modelling of ore flow in a belt conveyor system with cyclic loading. *Int. J. Min. Reclam. Environ.* **2018**, *32*, 376–391. [CrossRef]
26. Gładysiewicz, L. *Przenośniki Taśmowe Teoria i Obliczenia*; Oficyna Wydawnicza Politechniki Wrocławskiej: Wrocław, Poland, 2003.
27. Grima, A.P.; Wypych, P.W. Investigation into calibration of discrete element model parameters for scale-up and validation of particle-structure interactions under impact conditions. *Powder Technol.* **2011**, *212*, 198–209. [CrossRef]
28. Doroszuk, B.; Król, R. Analysis of conveyor belt wear caused by material acceleration in transfer stations. *Min. Sci.* **2019**, *26*, 189–201. [CrossRef]
29. Walker, P.; Doroszuk, B.; Król, R. Analysis of ore flow through longitudinal belt conveyor transfer point. *Eksploat. I Niezawodn.* **2020**, *22*, 536–543. [CrossRef]
30. Roessler, T.; Richter, C.; Katterfeld, A.; Will, F. Development of a standard calibration procedure for the DEM parameters of cohesionless bulk materials—Part I: Solving the problem of ambiguous parameter combinations. *Powder Technol.* **2019**, *343*, 803–812. [CrossRef]
31. Persson, B.N.J.; Tartaglino, U.; Albohr, O.; Tosatti, E. Rubber friction on wet and dry road surfaces: The sealing effect. *Phys. Rev.* **2005**, *71*. [CrossRef]
32. Roberts, A.W. *Characterisation for Hopper and Stockpile Design*; Blackwell Publishing Ltd.: Hoboken, NJ, USA, 2009; pp. 85–131. [CrossRef]
33. Kessler, F.; Prenner, M. DEM—Simulation of conveyor transfer chutes. *FME Trans.* **2009**, *37*, 185–192.
34. Roberts, A.W. Chute performance and design for rapid flow conditions. *Chem. Eng. Technol.* **2003**, *26*, 163–170. [CrossRef]

Article

Statistical Analysis and Neural Network in Detecting Steel Cord Failures in Conveyor Belts

Dominika Olchówka [1], **Aleksandra Rzeszowska** [2,*], **Leszek Jurdziak** [1] **and Ryszard Błażej** [1]

[1] Faculty of Geoengineering Mining and Geology, Wroclaw University of Science and Technology, Na Grobli 15, 50-421 Wroclaw, Poland; dominika.olchowka@pwr.edu.pl (D.O.); leszek.jurdziak@pwr.edu.pl (L.J.); ryszard.blazej@pwr.edu.pl (R.B.)

[2] Faculty of Electronics, Wroclaw University of Science and Technology, Janiszewskiego 11/17, 50-372 Wroclaw, Poland

* Correspondence: aleksandra.rzeszowska@pwr.edu.pl

Abstract: This paper presents the identification and classification of steel cord failures in the conveyor belt core based on an analysis of a two-dimensional image of magnetic field changes recorded using the Diagbelt system around scanned failures in the test belt. The obtained set of identified changes in images, obtained for numerous parameters settings of the device, were the base for statistical analysis. This analysis makes it possible to determine the Pearson's linear correlation coefficient between the parameters being changed and the image of the failures. In the second stage of the research, artificial intelligence methods were applied to construct a multilayer neural network (MLP) and to teach it appropriate identification of damage. In both methods, the same data sets were used, which made it possible to compare methods.

Keywords: conveyor belts; magnetic method; diagnostics; NDT method; belt damage; statistical analysis; neural networks

Citation: Olchówka, D.; Rzeszowska, A.; Jurdziak, L.; Błażej, R. Statistical Analysis and Neural Network in Detecting Steel Cord Failures in Conveyor Belts. *Energies* **2021**, *14*, 3081. https://doi.org/10.3390/en14113081

Academic Editors: Daniela Marasová, Monika Hardygora and Mirosław Bajda

Received: 6 May 2021
Accepted: 24 May 2021
Published: 26 May 2021

1. Introduction

The non-destructive testing (NDT) of conveyor belts gives vast possibilities related to the optimization of conveyor belt maintenance costs, such as choosing the right moment for repair, replacement or recondition of the belt based on, among other factors, detected damage [1] or the rate of change of belt thickness [2]. Belt operating time depends on several factors that are presented in the literature [2]. Among other things, it is affected by the hardness, size and shape of transported materials, the specificity of transport point and the length and age of belt cord. Some of these factors damage the belt covers or belt core. Figure 1 shows the cross-section of the example conveyor belt with steel cords.

NDT research use offers, inter alia, analysis of the magnetic field changes generated by damaged or missing cords. Research using this method has been carried out since 1970 [3]. Since then, several researchers around the world have developed various systems to detect damage to steel cords in the belt core [4–6]. One of the systems in use for studies is a Diagbelt magnetic system, which enables researchers to obtain two-dimensional images suitable for further analysis [7–9]. The device detects magnetic field changes arising during the movement of cord failures beneath the measuring probe (installed across the width of the belt) which generate a discredited signal (-1, 0 or 1) corresponding to measured values of the magnetic field. The positive value of the signal measured by the device is represented in the figures presented in this paper in blue, and the signal of the negative magnetic field is represented in yellow.

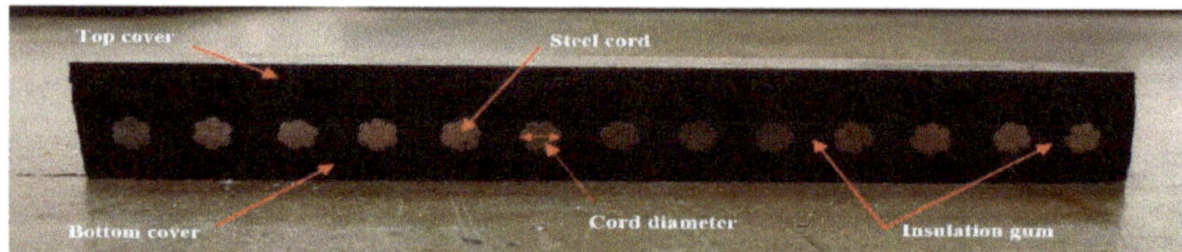

Figure 1. The cross-section of the belt with steel cords.

For the performed comparative analysis, a reference conveyor belt containing several artificial cord failures was used. The measurements were carried out by modifying system parameters: belt speed, the distance between the measuring probe and the cord and measurement sensitivity. These parameters were selected based on previous studies [10], which have confirmed their impact on received signals. The measurements were performed for many combinations of these parameters over ten measuring cycles, of which three were selected and used for the analysis. Preliminary visual evaluation of the data indicates the relationship between failure detection signal and the above parameters. Some of the damage with the appropriate settings of parameter values generated similar signals. Figure 2 gives an example of this situation (damage 3 and damage 1).

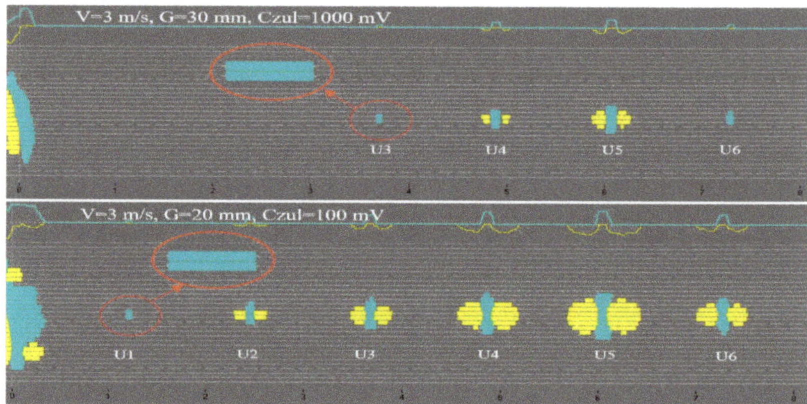

Figure 2. 2D images of the damage for significantly different settings of measuring equipment parameters.

The tested failures were divided into six categories: partial cord damage (20% (U1) and 50% (U2)) in one cord, complete cut of one steel cord (U3), cut of three (U4) and six (U5) cords and resection of one cord to the length of 20 mm (U6) (Figure 3).

2. Preparation of Data for Analysis

Each one of the cord failures described above generated a magnetic signal, and data were consolidated into 12 values, four for each of the sub-areas: magnetic signal surface areas, number of channels on which signal be detected, width and length of the signals. The method of calculating the size of signal for exemplary damage is shown in Figure 4 and described by Equations (1)–(3).

$$Z1sum = Z1\_1 + Z1\_2 + Z1\_3 + Z1\_4 \tag{1}$$

$$Nsum = N1 + N2 + N3 + N4 + N5 + N6 \tag{2}$$

$$Z2sum = Z2\_1 + Z2\_2 + Z2\_3 + Z2\_4 \tag{3}$$

where: Z1_klength of the signal detected on the k-th channel for the signal before damage, Nklength of the signal detected on the k-th channel for the signal of damage and Z2_klength of the signal detected on the k-th channel for the signal behind damage.

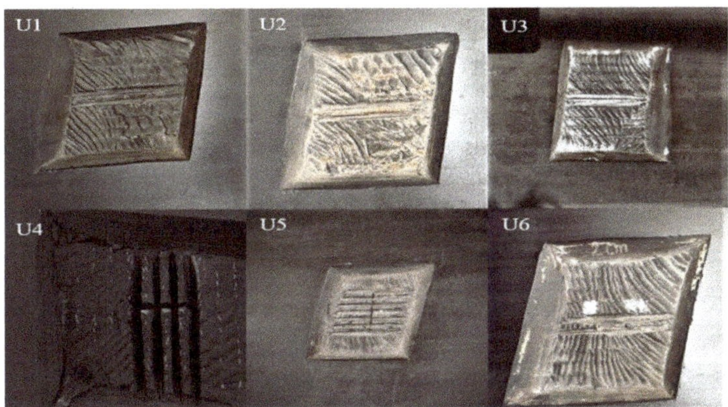

Figure 3. The actual appearance of the damage.

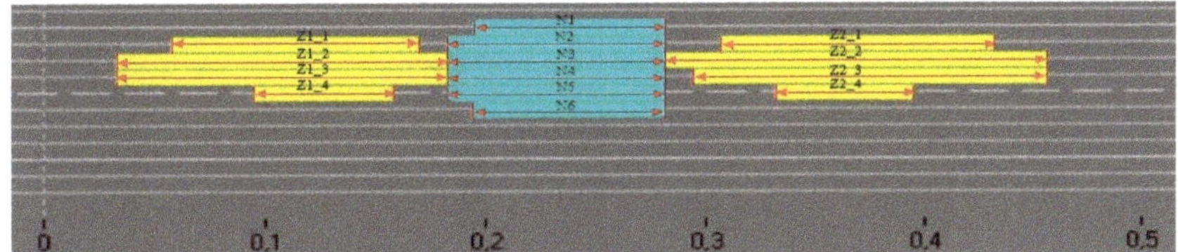

Figure 4. The method of calculating values describing selected damage.

During the measurements, the belt speed (V) was increased from 2 up to 5 m/s (in increases of 1 m/s), the distance between the measuring probe and the cord (g) was changed within the range of 20–50 mm (in increases of 10 mm) and the following sensitivity levels (c) were applied: 100 mV, 150 mV, 200 mV, 250 mV, 300 mV, 400 mV, 500 mV, 600 mV, 700 mV and 1000 mV. The selection of these parameters was predicated on technical capabilities (speed of the test conveyor) and also the observation of system behaviour and its settings in numerous previous studies [10]. The above parameter values are shown on the axis in Figure 5.

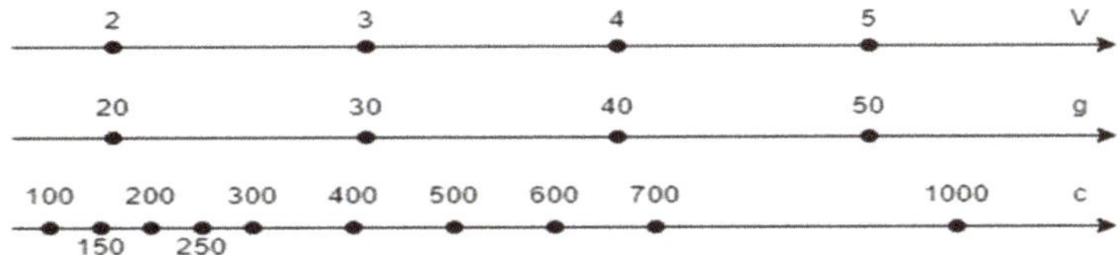

Figure 5. The distribution of measurement parameters.

The number of tested triple variants for given settings of the measuring system amounted to 160.

$$p = n_c \cdot n_v \cdot n_g = 4 \cdot 10 \cdot 4 = 160 \tag{4}$$

where: p—quantity of triples parameters variants, n_c—number of settings of sensitivity parameter, n_v—number of settings of belt speed and n_g—number of settings of distance between the measuring probe and the cord.

For each of the three measuring cycles for six defined types of damage, 2880 records describing the damage should be obtained.

$$L_p = p \cdot l_c \cdot l_k = 160 \cdot 3 \cdot 6 = 2880 \tag{5}$$

where: L_p—theoretical number of records, l_c—number of measuring cycles taken into account and l_k—number of types of damage.

The actual amount of data was lower (2367), because magnetic field changes were not detected for less core damage in certain measurement settings.

One paper [11] defines the most appropriate measuring system parameters, which are presented in Table 1. For these ranges' apparatus settings (three parameters), the number of output data sets decreased to the value of:

$$L_p = n_c \cdot n_v \cdot n_g \cdot l_c \cdot l_k = 4 \cdot 1 \cdot (4 + 3 + 3 + 2) \cdot 3 \cdot 6 = 864 \tag{6}$$

Table 1. Preferred measurement system parameter settings.

Belt Speed [m/s]	Range between the Belt Core and the Measuring Probe [mm]	Sensitivity [mV]
2	20–50	200–300
3	20–50	300–400
4	20–50	400–500
5	20–50	600–700

In reality, however, the number of records was 693 (some minor defects were not detected in specific measurement settings).

The actual sensitivities of the measuring device are inversely proportional to the parameter value called "sensitivity". When the value of this parameter is very low (e.g., 50–100 mV), the measuring system is extremely sensitive to the slightest field changes; however, the signal produced by the device is difficult to interpret. Images of the failures fuse and also appear to measure noise (Figure 6). Furthermore, when the value of this parameter is too large (i.e., when the system was set to be insensitive), minor damage may not have been registered, since it generates slight field changes which are outside the scope of sensitivity of the equipment.

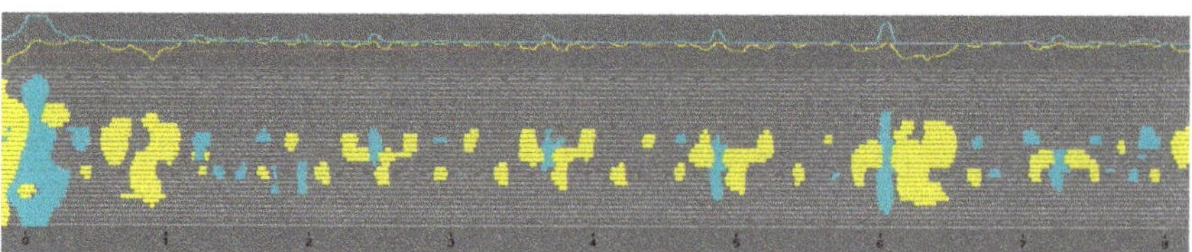

Figure 6. Inconclusive image of the damage for the measurement with high sensitivity of the measuring probe (scale 5 mm/pixel).

3. Statistical Analysis

The statistical analysis was performed only for the data obtained from the optimal sets of parameters (Table 2). The statistical analysis started with the verification of the obtained data to remove gross errors that may have appeared in the database, resulting, for example, from human oversight (e.g., entering incorrect data). In the next step, the correlation between the parameters taken into account in the analysis was examined. There are 13 such analysed values. They include damage number (Nr_U), number of cut cords (LL), area of damage (Pole_R), three parameters connected with the measurement system (belt speed—V, measuring probe distance—G, sensitivity—Czul), the measurement cycle taken into account (Cycle) and the failure description, including two values for each of the three damage sub-areas (yellow field before damage—Z1, blue field—N and yellow field behind the damage Z2). The measured damage parameters are the sum of the lengths of the signals recorded in the measurement channels (Z1sum, Nsum, Z2sum) and the number of channels recording the signal related to a given sub-area (Z1_LK, N_LK, Z2_LK).

Table 2. Confidence intervals for the mean of the measured parameters.

		U1	U2	U3	U4	U5	U6
Z1sum	$\overline{x}$	0.00	3.09	45.72	373.48	793.65	130.91
	Δ	0.00	12.67	44.37	94.09	136.12	80.08
Z1_LK	$\overline{x}$	0.00	0.09	0.79	3.40	5.17	1.61
	Δ	0.00	0.08	0.23	0.17	0.21	0.27
Nsum	$\overline{x}$	20.50	110.20	231.10	542.26	821.79	334.82
	Δ	2.07	8.29	13.12	16.09	25.25	14.44
N_LK	$\overline{x}$	0.88	2.29	3.68	6.01	8.04	4.46
	Δ	0.07	0.12	0.14	0.11	0.14	0.15
Z2sum	$\overline{x}$	0.00	2.40	43.95	395.60	848.97	103.40
	Δ	0.00	3.83	14.06	32.92	45.95	21.77
Z2_LK	$\overline{x}$	0.00	0.04	0.61	3.24	4.89	1.24
	Δ	0.00	0.05	0.18	0.19	0.18	0.23

Figure 7 presents charts showing the values of six measured parameters for damage depending on the class to which the damage belongs. A visual evaluation of the data helps to decide whether a given parameter affects the class differentiation or is irrelevant and can be removed.

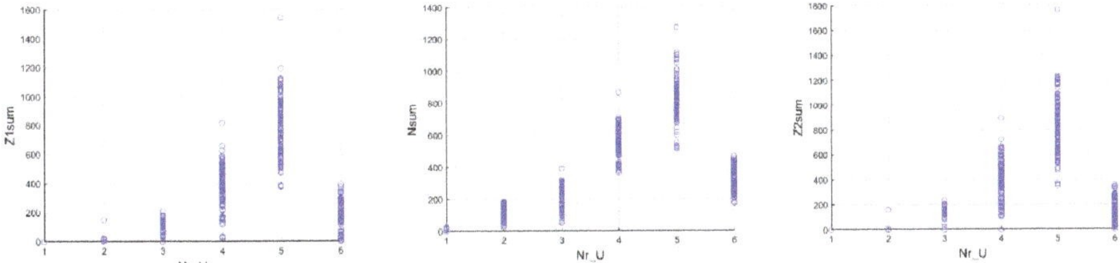

Figure 7. Distribution of the values of Z1sum, Nsum and Z2sum depending on the damage class.

The first part of the statistical analysis determined confidence intervals for the mean for each of the analysed measurements [12]. Confidence intervals for the mean are given as a formula:

$$\overline{x} \pm \Delta = \overline{x} \pm 1.96 \frac{\sigma}{\sqrt{N}} \tag{7}$$

where: $\overline{x}$—sample mean, Δ half the width of the confidence interval, σ standard deviation and N number of samples.

The input base is divided into two parts: a training set and a test set. Every third value went to the test set, while the remaining samples were left in the training set. The size of the training set was 462 samples, and the size of the test set was 231.

Table 2 summarizes the calculated values that facilitate the determination of the confidence interval for each of the analysed data sets.

For the data from the test, the set was determined and the mean value of each of the test sets was prepared in this way. These values were placed on the graph, which also marks the widths of the determined confidence intervals (Figure 8). Table 3 summarizes the results obtained from a given test group. In the table, the values that fall outside the designated confidence interval are marked in red.

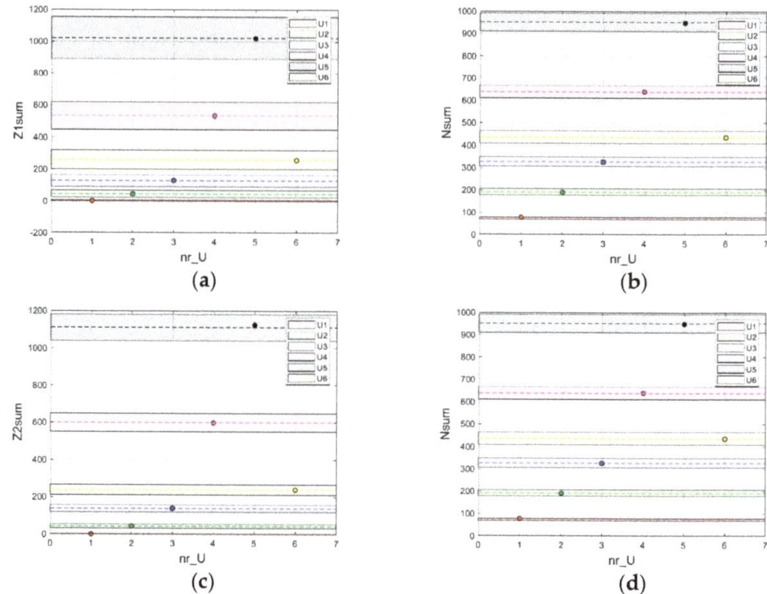

Figure 8. Confidence intervals test set, (**a**)–Z1sum (overlapping intervals), (**b**)–Nsum, (**c**)–Z2sum, (**d**)–N_LK (data outside of assigned value).

Table 3. The mean values of the test sets.

	U1	U2	U3	U4	U5	U6
Z1sum	0.00	1.03	45.79	377.27	788.06	131.00
Z1_LK	0.00	0.06	0.75	3.40	5.17	1.63
Nsum	20.75	107.94	230.48	538.98	815.54	332.23
N_LK	1.00	2.26	3.67	6.04	8.04	4.46
Z2sum	0.00	0.00	42.33	400.04	843.58	104.23
Z2_LK	0.00	0.00	0.60	3.19	4.92	1.21

Therefore, it can be noticed that the problem with recognition appears only in the case of data concerning the first type of damage. The number of channels in the test sample was mean 1.00, and in the training sample was 0.88 ± 0.07.

To check the influence of the analysed values on each other (their linear association), a statistical analysis was performed which determined the Pearson's product–moment linear correlation coefficients between the setting parameters of the measurement device and the sizes of output signals for each of the analysed failures. Pearson's correlation coefficient is a measure of linear correlation between two sets of data (the covariance of the two variables divided by the product of their standard deviations). Figure 9 displays the estimated correlations in the form of a matrix with coloured cells. Small changes to controlled parameters and lack of outliers in results allowed us to assume the linearity of changes in results. It was applied as an initial test of linear associations for further investigations to find the physical influence of settings changes on results, to select the best settings for given working conditions of conveyor belts and to select appropriate output parameters and methods for steel cord belt failures classification.

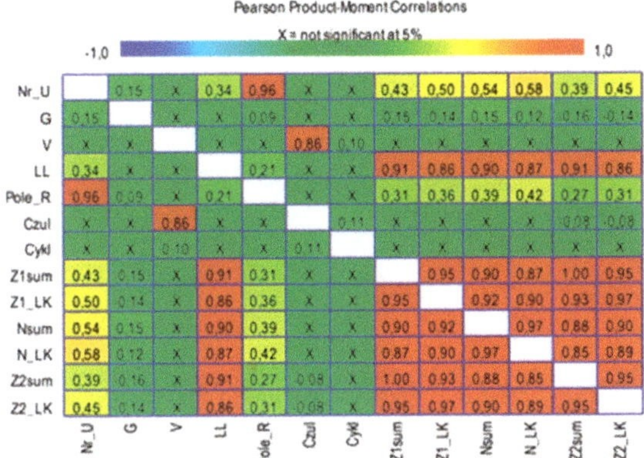

Figure 9. Pearson's linear correlation coefficients.

The data in the table above (Figure 9) are displayed using both colours and numerical values. Data marked with "X" are statistically insignificant. Coefficients define a linear relationship between two different variables. The greater the value of the correlation parameter, the greater the degree of interconnection between the pairs of variables. It is worth noting that the presented table shows that the selected parameters of the measuring system did not significantly affect the type of damage. No correlation was found between the belt speed and the measurement results (statistically insignificant correlation), there is a low correlation between the measuring head distance and the measurement results (negative or positive, depending on the area, within the range -0.16–0.15 and no significant correlation or weak negative correlation (-0.08) of the measurement results to the sensitivity of the device.

It is also worth noting that all measurement results are strongly positively correlated with each other, and the correlation between the values describing the yellow fields (Z1sum and Z2sum and Z1_LK and Z2_LK) is 1.00 and 0.97, which maintains the hypothesis about their symmetry [1].

Analogical statistical analysis for the full data set was described in detail in [13], but the results obtained there turned out to be less satisfactory than the results obtained for specific parameters of the measurement system. The distribution of the data from the complete set is shown in Figure 10 (these graphs show the values of Z1sum, Nsum and N_LK). These results largely overlap, and it is impossible to clearly define the boundaries of clusters [14,15].

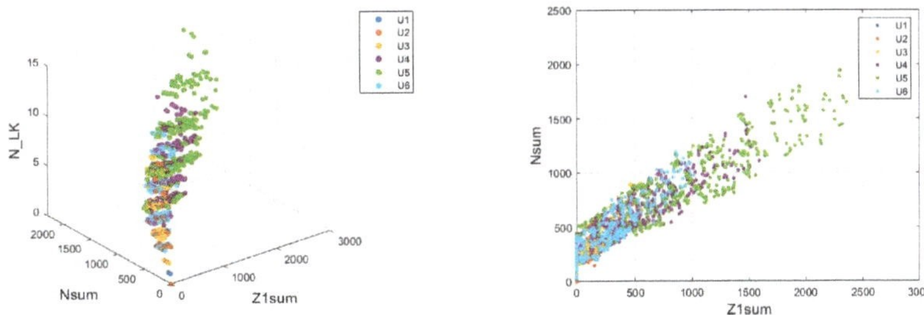

Figure 10. Cluster analysis—full data set, dependence Z1sum, Nsum, N_LK.

Similar graphs (Figure 11) were also plotted for the set of parameters of the measurement system tested in this analysis. It can be noticed that, in this case, it is possible to limit the obtained data with a certain curve marking the boundary of a given cluster, although there are still areas where the belonging of the measurement result to a given cluster is ambiguous.

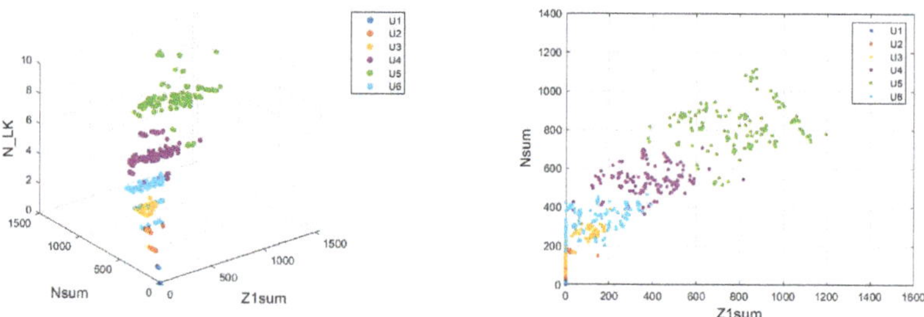

Figure 11. Cluster analysis—limited data set, dependence Z1sum, Nsum, N_LK.

Optical analysis of the obtained charts shows that the measurement data are highly probable to correctly classify the type of damage, but there are areas where the classification may fail because clusters overlap [15]. Due to this fact, another analysis was carried out using artificial neural networks.

4. Analysis with the Use of Neural Networks

The analysis of the selection of the structure and parameters of the neural network as well as the idea of its operation has been widely described in the literature. The studies [16–18] describe in detail the rationale behind the selection of specific parameters used in this research. The MATLAB software with the Deep Learning Toolbox installed was used for the learning process of neural networks. The way of using this toolbox is described in studies [19,20]. This toolbox enables the creation of a multilayer neural network with a given number of neurons in each of the hidden layers, and the ability to train the neural network on a given training data set (which is automatically divided into a training and validation set). Each layer of the artificial neural network is made up of individual computing units called neurons. Neurons, stimulated by the signal fed to their input, work out the output signal using the assigned weight (w) and the added bias (b). The process of training a neural network under supervision consists of repeatedly assessing it in the process of training samples from the test set, and then updating the weight based on the error between the network response and the expected response. To use artificial neural networks to classify the conveyor belt damage, it was necessary to generate appropriate sets of training and test sets, train the network on the training set, and then test it on the test set. Since neural networks can divide the classification of space non-linearly, over the course of this research two variants of the selection and division of the input data were distinguished. In each case, the vector of input data consisted of 15 elements (and this is the number of neurons in the input layer of the neural network): 3 measurement parameters and 4 values describing each of the three sub-areas. There are 6 neurons in the output layer—one each responsible for belonging to a given class of damage. Two hidden layers consisting of 31 and 63 neurons were placed between the input and output layers. The size of these layers was determined by the Kolmogorov theorem, according to which the number of neurons in the hidden layer should be equal to the number of neurons in the previous layer multiplied by two and increased by one [21]. The diagram of the neural network used in this study is shown in Figure 12.

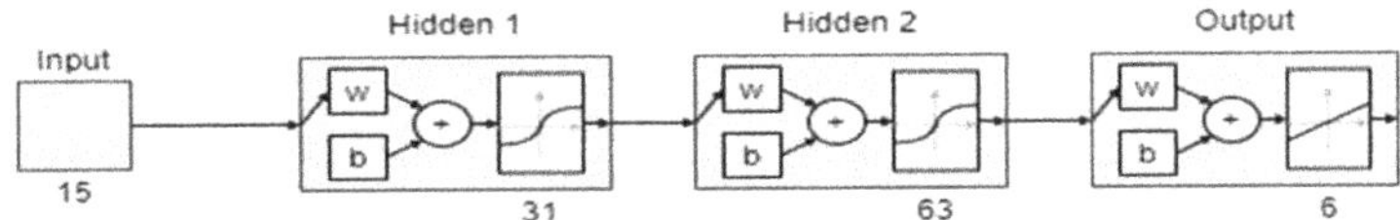

Figure 12. Diagram of the neural network used in the research.

Two variants were used for the process of learning neural networks:

- variant 1: analysis of all available data, two measurement cycles of each parameter set go to the training set, one cycle to the test set. The number of training and test sets was 1578 and 789.
- variant 2: analysis of data obtained in the measurements with the best possible sets of parameters (Table 1), two measurement cycles of each of the included parameter sets go to the training set, one cycle to the test set. The number of training and test sets was 462 and 231.

Each of the network learning processes in a given variant was performed three times, and the results are presented in Table 4. The effectiveness of the diagnosis was divided depending on the type of damage that the neural network was supposed to recognize. In addition, the effectiveness of the diagnosis was also determined for the entire test set.

Table 4. Measurement results of neural networks analysis.

Variant	No	Recognition Effectiveness [%]						
		U1	U2	U3	U4	U5	U6	Total
	1	98.57	99.19	99.67	99.69	100.00	100.00	99.37
1	2	97.14	97.56	98.68	98.13	100.00	100.00	98.86
	3	100.00	98.37	98.03	98.13	100.00	100.00	98.99
	1	100.00	97.14	97.92	100.00	100.00	100.00	99.13
2	2	100.00	97.14	97.92	100.00	100.00	100.00	99.13
	3	100.00	100.00	100.00	100.00	100.00	100.00	100.00

5. Conclusions

Many scans performed over the years using the Diagbelt system have shown that the magnetic measuring system is well suited to obtaining detailed information about the technical condition of the belt core. The idea of the system is based on the measurement of magnetic field changes at sites of core damage. The data obtained are presented on two-dimensional pictures that can be easily analysed using proposed known methodologies. The results presented in this study confirm this observation, showing the high efficiency of identification of the type of damage using the signals generated by Diagbelt.

One such method is verification based on the average of the samples. For this purpose, from the set of training data the mean value and 95% confidence interval were calculated for each damage. Then, for the test set containing the specific damage data obtained at different set values (not taken for training), the mean value was determined, which allowed for its comparison with the previously determined confidence intervals. In nearly every instance, testing set data have been included in the relevant confidence interval. The number of samples in training and testing data for selected parameter sets was 8 and 4, respectively, which could be a too-small value. It is, however, worth noting that with such a choice of analysis for automatic recognition of damage, it is necessary to execute many damage measurements for multiple sets of parameters to obtain a sample that comes from the same distribution as the training data. This solution can be cumbersome and, as the study shows, verification by the mean of the sample data is not always reliable.

An analysis of the Pearson correlation coefficient allows an assessment of the interdependence of evaluated parameters and therefore initially verifies which parameters are worth analysing with the classification of measuring damage, and which are redundant and have no correlation with the type of damage. Such an analysis does not allow for the designated similar data based on a new sample, but on its basis it is possible to construct a statistical model necessary for the assessment of future data. Creating such a model, which is a response to the analysis presented in this article, is a good direction for future research.

When the full set of data is limited to data obtained for the most appropriate system settings, better statistical analysis values are achieved. The full data set shown in the two-dimensional and three-dimensional plots (Figures 9 and 10) indicates that these data cannot be isolated from each other and locked in separate clusters; however, for limited data, cluster analysis is possible, as areas of interdependent neighbouring clusters are slight.

The analysis based on neural networks allows the omission of the problem of non-linearity. The network containing two hidden layers allows investigators to solve almost every problem of classification, provided it has the appropriate input data. In the framework of this explanation, analysis was carried out using neural networks, both on the complete dataset and the limited dataset containing results obtained from best possible system settings. The data collected in both of these variants have shown good efficacy (above 98%) following the implementation of the testing process.

The network does not have a problem with classification of the last two types of damage (U5 and U6); however, it makes errors recognising defects 2–4. This may be due to the real size of the defect concerned. The neural network analysis, compared to the statistical analysis, allows for quick action of the entire system while maintaining high efficiency.

It is worth noting that while analysis of statistical methods has already been used in the classification of belts damage, cluster analysis and analysis using the neural networks have so far been rarely discussed and their results rarely presented. Developers of the diagnostic systems, in many cases, prefer to retain the ability to interpret measurement results for themselves, so that their services will not become unnecessary. In the Industry 4.0 era [22], the automatic interpretation of the diagnostic signal is necessary to cope with data processing for an ever-increasing amount of data. Test results discussed in this paper are promising, and they show the direction of further action that authors are taking as part of the research project "Integrated mobile system of automatic testing and continuous diagnostics of the condition of conveyor belts" (project number: POIR.01.01.01-00-1194/19) [23].

Author Contributions: Conceptualization, R.B. and L.J.; methodology, D.O., A.R. and L.J.; formal analysis, A.R. and L.J.; resources, R.B.; data curation, D.O., A.R.; writing—original draft preparation, D.O. and A.R.; writing—review and editing, R.B and L.J.; supervision, R.B. All authors have read and agreed to the published version of the manuscript.

Funding: This research received no external funding.

Institutional Review Board Statement: Not applicable.

Informed Consent Statement: Not applicable.

Data Availability Statement: The data presented in this study are available on request from the corresponding author.

Conflicts of Interest: The authors declare no conflict of interest.

References

1. Jurdziak, L.; Błażej, R.; Burduk, A.; Bajda, M.; Kirjanów-Błażej, A.; Kozłowski, T.; Olchówka, D. Optimization of the operating costs of conveyor beltsin a mine in various strategies of their replacement and diagnostics. *Transp. Przemysłowy Masz. Rob.* **2020**, *4*, 14–22.
2. Webb, C.; Sikorska, J.; Khan, R.N.; Hodkiewicz, M. Developing and evaluating predictive conveyor belt wear models. *Data-Cent. Eng.* **2020**, *1*, e3. [CrossRef]
3. Barry, B. Multi-Channel Conveyor Belt Condition Monitoring. U.S. Patent 7,275,637, 2 October 2007.
4. Błażej, R. *Assessment of the Technical Condition of Conveyor Belts with Steel Cords*; Oficyna Wydawnicza Politechniki Wrocławskiej: Wrocław, Poland, 2018. (In Polish)

5.	Kwaśniewski, J. *Magnetic Testing of Steel Ropes—Personnel Certification System in the MTR Method*; Wydawnictwa AGH: Kraków, Poland, 2010. (In Polish)
6.	Kwaśniewski, J.; Machula, T. Diagnostics of belts with steel cords. *Transp. Przemysłowy* **2006**, *4*, 22–24. (In Polish)
7.	Błażej, R.; Jurdziak, L.; Kirjanów, A.; Kozłowski, T. Evaluation of the quality of steel cord belt splices based on belt condition examination using magnetic techniques. *Diagnostyka* **2015**, *16*, 59–64.
8.	Błażej, R.; Jurdziak, L.; Kozłowski, T.; Kirjanów, A. The use of magnetic sensors in monitoring the condition of the core in steel cord conveyor belts—Tests of the measuring probe and the design of the DiagBelt system. *Measurement* **2018**, *123*, 48–53. [CrossRef]
9.	Kirjanów, A.; Burduk, R. Regression method-based analysis of damage development in the core of steel cord conveyor belts. In Proceedings of the XVIII Conference of PhD Students and Young Scientists, Szklarska Poręba, Poland, 22–25 May 2018; p. 5.
10.	Kozłowski, T.; Błażej, R. 2018 Influence of selected parameters on the magnetic testing results in a DiagBelt system. *Transp. Przemysłowy Masz. Rob.* **2018**, *4*, 6–9. (In Polish)
11.	Olchówka, D. Selection of Measurement Parameters Using the DiagBelt Magnetic System on the Test Conveyor. Master's Thesis, Wroclaw University of Science and Technology, Wroclaw, Poland, 2020. (In Polish)
12.	Hazra, A. Using the confidence interval confidently. *J. Thorac. Dis.* **2017**, *9*, 4125–4130. [CrossRef] [PubMed]
13.	Olchówka, D.; Jurdziak, L.; Błażej, R.; Kozłowski, T. Examination of the steel cord belts core damage condition using the DiagBelt system Part 1. Determining the number of steel cord cuts in damage. *Transp. Przemysłowy Masz. Rob.* **2021**, *1*, 2–10. (In Polish)
14.	Omran, M.; Engelbrecht, A.; Salman, A. An overview of clustering methods. *Intell. Data Anal.* **2007**, *11*, 583–605. [CrossRef]
15.	Saxena, A.; Prasad, M.; Gupta, A.; Bharill, N.; Patel op Tiwari, A.; Er, M.; Lin, C. A Review of Clustering Techniques and Developments. *Neurocomputing* **2017**, *267*, 664–681. [CrossRef]
16.	Hornik, K.; Stinchcombe, M.; White, H. Multilayer feedforward networks are universal approximators. *Neural Netw.* **1989**, *2*, 359–366. [CrossRef]
17.	Osowski, S. *Neural Networks for Information Processing*; Oficyna Wydawnicza Politechniki Warszawskiej: Warszawa, Poland, 2020. (In Polish)
18.	Phil, K. *MATLAB Deep Learning: With Machine Learning, Neural Networks and Artificial Intelligence*; Apress: Berkeley, CA, USA, 2017.
19.	Anghel, D.; Ene, A.; Belu, N. A Matlab Neural Network Application for the Study of Working Conditions. *Adv. Mater. Res.* **2013**, *837*, 310–315. [CrossRef]
20.	Demuth, H.; Beale, M. *Neural Network Toolbox User's Guide*; MathWorks, Inc.: Portola Valley, CA, USA, 2014.
21.	Osowski, S. *Neural Networks in Algorithmic Use*; Wydawnictwo Naukowo Techniczne: Warszawa, Poland, 1996. (In Polish)
22.	Jurdziak, L.; Błażej, R.; Bajda, M. Conveyor belt 4.0, Intelligent systems in production engineering and maintenance. *Adv. Intell. Syst. Comput.* **2019**, *835*, 645–654.
23.	Integrated Mobile System of Automatic Testing and Continuous Diagnostics of the Condition of Conveyor Belts. Available online: https://mapadotacji.gov.pl/projekty/1179135/ (accessed on 30 April 2021).

MDPI
St. Alban-Anlage 66
4052 Basel
Switzerland
Tel. +41 61 683 77 34
Fax +41 61 302 89 18
www.mdpi.com

Energies Editorial Office
E-mail: energies@mdpi.com
www.mdpi.com/journal/energies